Bionanodesign
Old Forms for New Functions
2nd Edition

Nanoscience & Nanotechnology Series

Editor-in-chief:
Nguyễn T. K. Thanh, *University College London, UK*

Series editors:
Gabriel Caruntu, *Central Michigan University, USA*
Shinya Maenosono, *Japan Advanced Institute of Science and Technology, Japan*
Neerish Revaprasadu, *University of Zululand, South Africa*

Titles in the series:
1: Nanotubes and Nanowires
2: Fullerenes: Principles and Applications
3: Nanocharacterisation
4: Atom Resolved Surface Reactions: Nanocatalysis
5: Biomimetic Nanoceramics in Clinical Use: From Materials to Applications
6: Nanofluidics: Nanoscience and Nanotechnology
7: Bionanodesign: Following Nature's Touch
8: Nano-Society: Pushing the Boundaries of Technology
9: Polymer-based Nanostructures: Medical Applications
10: Metallic and Molecular Interactions in Nanometer Layers, Pores and Particles: New Findings at the Yoctolitre Level
11: Nanocasting: A Versatile Strategy for Creating Nanostructured Porous Materials
12: Titanate and Titania Nanotubes: Synthesis, Properties and Applications
13: Raman Spectroscopy, Fullerenes and Nanotechnology
14: Nanotechnologies in Food
15: Unravelling Single Cell Genomics: Micro and Nanotools
16: Polymer Nanocomposites by Emulsion and Suspension
17: Phage Nanobiotechnology
18: Nanotubes and Nanowires, 2nd Edition
19: Nanostructured Catalysts: Transition Metal Oxides
20: Fullerenes: Principles and Applications, 2nd Edition
21: Biological Interactions with Surface Charge Biomaterials
22: Nanoporous Gold: From an Ancient Technology to a High-Tech Material
23: Nanoparticles in Anti-Microbial Materials: Use and Characterisation
24: Manipulation of Nanoscale Materials: An Introduction to Nanoarchitectonics
25: Towards Efficient Designing of Safe Nanomaterials: Innovative Merge of Computational Approaches and Experimental Techniques
26: Polymer–Graphene Nanocomposites
27: Carbon Nanotube–Polymer Composites
28: Nanoscience for the Conservation of Works of Art
29: Polymer Nanofibers: Building Blocks for Nanotechnology
30: Artificial Cilia

31: Nanodiamond
32: Nanofabrication and its Application in Renewable Energy
33: Semiconductor Quantum Dots: Organometallic and Inorganic Synthesis
34: Soft Nanoparticles for Biomedical Applications
35: Hierarchical Nanostructures for Energy Devices
36: Microfluidics for Medical Applications
37: Nanocharacterisation, 2nd Edition
38: Thermometry at the Nanoscale: Techniques and Selected Applications
39: Nanoceramics in Clinical Use: From Materials to Applications, 2nd Edition
40: Near-infrared Nanomaterials: Preparation, Bioimaging and Therapy Applications
41: Nanofluidics, 2nd Edition
42: Nanotechnologies in Food, 2nd Edition
43: ZnO Nanostructures: Fabrication and Applications
44: Diatom Nanotechnology: Progress and Emerging Applications
45: Nanostructured Materials for Type III Photovoltaics
46: Chemically Derived Graphene: Functionalization, Properties and Applications
47: Graphene-based Membranes for Mass Transport Applications
48: Carbon Nanostructures for Biomedical Applications
49: Surface Chemistry of Colloidal Nanocrystals
50: Reducing Agents in Colloidal Nanoparticle Synthesis
51: Carbon Nitride Nanostructures for Sustainable Energy Production and Environmental Remediation
52: Nanotubes and Nanowires, 3rd Edition
53: Bionanodesign: Old Forms for New Functions, 2nd Edition

How to obtain future titles on publication:
A standing order plan is available for this series. A standing order will bring delivery of each new volume immediately on publication.

For further information please contact:
Book Sales Department, Royal Society of Chemistry, Thomas Graham House, Science Park, Milton Road, Cambridge, CB4 0WF, UK
Telephone: +44 (0)1223 420066, Fax: +44 (0)1223 420247
Email: booksales@rsc.org
Visit our website at www.rsc.org/books

Bionanodesign
Old Forms for New Functions
2nd Edition

By

Maxim Ryadnov
National Physical Laboratory, UK
Email: max.ryadnov@npl.co.uk

ROYAL SOCIETY OF **CHEMISTRY**

Nanoscience & Nanotechnology Series No. 53

Print ISBN: 978-1-78262-816-3
PDF ISBN: 978-1-78801-000-9
EPUB ISBN: 978-1-78801-931-6
Print ISSN: 1757-7136
Electronic ISSN: 1757-7144

A catalogue record for this book is available from the British Library

The Royal Society of Chemistry is a charity, registered in England and Wales, Number 207890, and a company incorporated in England by Royal Charter (Registered No. RC000524), registered office: Burlington House, Piccadilly, London W1J 0BA, UK, Telephone: +44 (0) 20 7437 8656.

For further information see our web site at www.rsc.org

Printed in the United Kingdom by CPI Group (UK) Ltd, Croydon, CR0 4YY, UK

Preface to the First Edition

The progress of today's science and technology encounters an increasing demand for finer and more efficiently performing materials with properties superior to those of current and hence ageing devices. Whether this concerns electronics or drug delivery, cancer diagnostics or alternative energy sources the search for means of miniaturising the existing materials or devising fundamentally new components with higher capacities appears to be relentless.

A saving solution to this is widely proposed as the design and fabrication of nanostructures, molecular architectures with dimensions featured below 100 nm.

By convention, and as originally formulated by Richard Feynman, the challenge of constructing macroscopic structures through the manipulation of individual molecules or even atoms prompted the emergence of a rapidly evolving field – nanotechnology. By definition, nanotechnology mirrors complex organisation at the nanoscale and is underpinned by a variety of related physical events that are combined into one universal process – molecular self-assembly.

The phenomenon of self-assembling molecules is attractive from both academic and application perspectives. However, preferential attention is being given to approaches whereby nanostructured materials or their components can be produced, moreover, produced at whim; that is, designed.

The pursuit for routes that can lead to rational or at least predictable design strategies invoked the main objective of this publication – to bring together contemporary approaches for designing nanostructures that employ naturally derived self-assembling motifs as synthetic platforms.

Nanoscience & Nanotechnology Series No. 53
Bionanodesign: Old Forms for New Functions, 2nd Edition
By Maxim Ryadnov

Published by the Royal Society of Chemistry, www.rsc.org

Entitled *bioinspired nanoscale design* or *bionanodesign* the book is written in the shape of a review, referenced as fully as permissible within the context of biomolecular recognition and self-assembly, which forms a general trend throughout.

The volume is composed of three core chapters focusing on three prominent topics of applied nanotechnology where the role of nanodesign is predominant. Specific applications that arise from designed nanoscale assemblies as well as fabrication and characterisation techniques are of a much lesser focus and whenever they appear serve as progress and innovation highlights.

In this sense, the book takes a nonstandard approach in delivering the material of this kind. It does not lead straight to applications or methods as most nanotechnology titles tend to do, but instead it admits the initial and primary stress on "nano" rather than on "technology". The task is significantly eased by the cohort of brilliant bioinspired designs reported to date and complicated by the volume they create almost on a weekly basis. For this reason, the author apologises for the inevitable, but not necessarily deliberate, omission of examples, many of which may prove to be equally if not more influential in bionanodesign.

Maxim Ryadnov

Preface to the Second Edition

Biomolecular structures self-assembled from naturally occurring motifs provide access to a wide variety of nanomaterials. Perfected in biology, such motifs support functions as diverse as molecular transport and tissue repair, while being used to construct nearly any conceivable nanomaterial form.

This book sets out to bring together contemporary approaches developed to purpose and re-purpose self-assembling motifs for nanomaterials with desired properties. The volume continues the course taken in the first edition, focusing on *bioinspired nanoscale designs*, but extends the scope of existing and emerging designs to functions which are atypical for well-established nanoscale shapes and forms. Therefore, emphasis is placed on adapting known or "old" forms, be these virus-like particles, DNA double helix or fibrillar matrices, for functions which are "new" to these forms.

The book samples most recent and background literature striving to be referenced as fully as permissible within the context of biologically relevant nanostructures and materials and comprises three core chapters, with each discussing one prominent topic of biological function defined at the nanoscale. Each chapter details structure–function relationships used to inform design rules for functional nanomaterials, giving examples of specific applications that inspire or were inspired by specific designs. Fabrication and characterization methodologies are also described to highlight progress in the field.

The book takes an unconventional approach in that it does not lead straight to applications or focus exclusively on the art of nanoscale design as most nanotechnology titles tend to do, but instead accepts the primary role of "nano" in defining "technology" leaning towards challenges where nano can introduce new solutions. This task finds support from the growing

Nanoscience & Nanotechnology Series No. 53
Bionanodesign: Old Forms for New Functions, 2nd Edition
By Maxim Ryadnov

Published by the Royal Society of Chemistry, www.rsc.org

number of excellent bioinspired nanoscale designs sourced in the literature and is continuously challenged by the volume they create almost on a weekly basis. For this reason, the author apologises for the inevitable, but not deliberate, omission of many examples which may prove to be equally if not more influential for bionanodesign.

Maxim Ryadnov

Contents

Nanoscience & Nanotechnology Series No. 53
Bionanodesign: Old Forms for New Functions, 2nd Edition
By Maxim Ryadnov

Published by the Royal Society of Chemistry, www.rsc.org

CHAPTER 1

Introductory Notes

The last decade has witnessed considerable progress in the development of technologies pertinent to the life sciences. This progress has encountered an increasing demand for finer materials with properties superior to those of current devices.[1] Performance efficacy, functional versatility and ease of use are amongst the major factors behind the choice of components and fabrication methodologies for next-generation materials. In the search for suitable and most effective approaches, molecular self-assembly has emerged as a strategy meeting all the current requirements.[2] As the progress continues, the strategy also reveals the pros and cons of technological developments, revising them and revealing new criteria for improvements. Perfected by Nature, the processes of molecular self-assembly are being successfully adapted by design providing reliable rules for the assembly of diverse architectures whose structure and function are defined at the nanoscale. Naturally occurring systems continue to be a major source of inspiration for self-assembly motifs and nanoscale designs. Principally, artificial designs are mimetics of native biology, aiming to devise desired functions with more effective outcomes. Cross-validation of motifs, which are unknown for one molecular class but common for another, also takes place in the design process, adding to the growing toolbox of construction components. The same can be said about functions. Each function has a precedent in natural systems, but not all systems can be readily adapted by design. Therefore, target functions can be atypical for chosen molecular forms. For example, a virus capsid is a recognizable form in biology, that is, an old form, but if designed to destroy bacterial cells on contact it delivers an atypical function, that is, a new function.[3] Another example is DNA. It is a universal, old, genetic material for all forms of life, but if assembled into a nanoscale vault as a reactor for enzymatic reactions,[4] it becomes assigned to an atypical, new function, Figure 1.1.

Nanoscience & Nanotechnology Series No. 53
Bionanodesign: Old Forms for New Functions, 2nd Edition
By Maxim Ryadnov

Published by the Royal Society of Chemistry, www.rsc.org

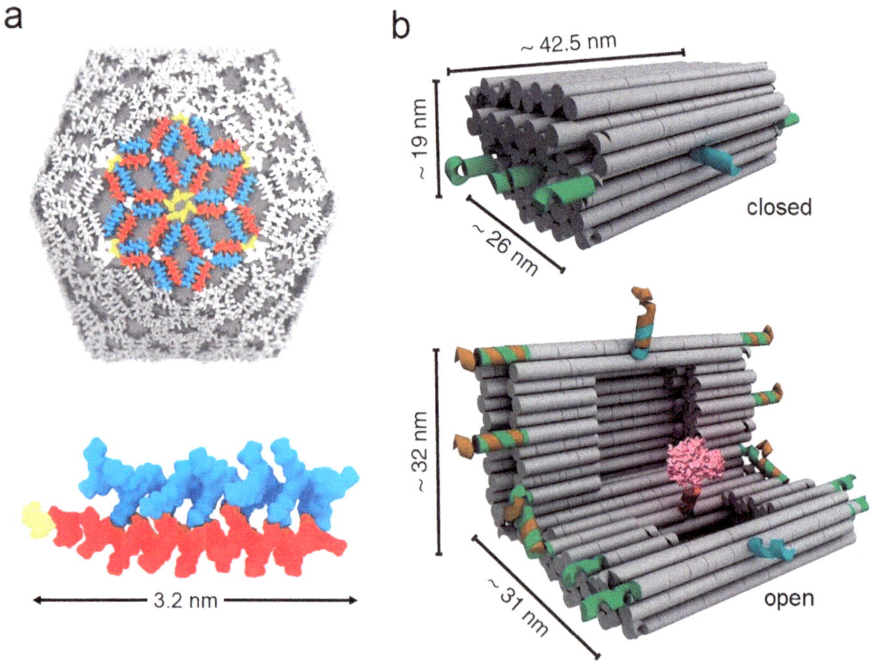

Figure 1.1 Old forms for new functions. (a) An artificial protein capsid designed to kill bacteria on contact by lysing their cell membranes. The capsid incorporates antimicrobial sequences (blue) and counterpart sequences (red) mediating the assembly. Reproduced from ref. 3, https://doi.org/10.1038/s41467-017-02475-3, under the terms of the CC BY 4.0 license https://creativecommons.org/licenses/by/4.0/. (b) A DNA assembly designed as a nanoscale vault hosting an enzyme molecule (pink) in the closed (top) and open (bottom) states. Reproduced from ref. 4, https://doi.org/10.1038/s41467-017-01072-8, under the terms of the CC BY 4.0 license https://creativecommons.org/licenses/by/4.0/.

This view conforms to the mission of bioinspired nanoscale design, or *bionanodesign*, which seeks to exploit engineering principles of biology to build up nanostructures, *i.e.* molecular architectures with regular periodicities featured below 100 nm. As formulated by Richard Feynman,[5] the challenge of understanding matter by constructing macroscopic structures from the bottom up prompted the field of nanotechnology. Self-assembling molecules lend themselves as primary components for nanostructures, and together with the rules of their assembly provide necessary tools towards the production of desired materials and away from alternative, dysfunctional forms.

The hierarchical nature of self-assembly is what reserves predictability and reproducibility in the assembly of resultant materials. In particular, this is the case for biopolymers, such as polypeptides and nucleic acids, which determine the morphological diversity of biological structures and can serve as ready-to-use design templates. However, additional constraints are also required to maximize the accuracy of a target nanoscale assembly from a chosen biopolymer type. This is addressed by the intrinsic propensity of self-assembly for autonomous control over supramolecular propagations of individual molecules.[6] These involve molecularly encoded folding, which correlates each hierarchical level with a self-assembly pattern, thus allowing biopolymer building blocks to assemble with a precision of a single nanometre.

Our ability to reproduce such a state of control and prediction has improved considerably over the last decade, although it is limited owing to an incomplete understanding of molecular self-assembly processes. More insight gained into biomolecular hierarchies will lead to new molecular models and designs with new properties, which may be perceived as unachievable today. Therefore, reviewing available guidance for the fabrication of functional nanostructures remains of paramount importance.

Replicating Nature's designs faithfully reproduced over millions of years has proven to be the most straightforward route to success. Nature offers a wealth of structural motifs capable of defining nanoscale forms, including DNA topologies, protein polyhedra, multicomponent extracellular matrices and biological membranes. Despite their diversity, all such motifs point to a robust design rationale for the broadest spectrum of possible nanostructures – but how to extract such a rationale for engineering artificial systems?

Of different types as well as within every type, Nature's designs are individually unique as required by functions that they carry or are assigned to. On the one hand, this underpins conserved templates adaptable for synthetic designs. On the other hand, biopolymers obey the same assembly principle: they adopt secondary structures to build functional quaternary systems, *i.e.* natural nanoscale objects. Synthetic designs reported to date take both routes. Protein or DNA structures based on pre-assembled native folds as well as systems designed from scratch, but inspired by native structures, are peers. Thus, a general approach to tackle the problem may focus on the assimilation of Nature's way in creating macromolecular assemblies by employing the structure–assembly relationships of existing examples. Eventually, this may reveal the essence of that structure-based strategy, which will allow us to exploit biomolecular recognition for nanoscale designs. Indeed, nanostructured systems shown as more advanced tend to assemble from better understood assembly elements. For instance, designs derived from DNA offer precision and control to match, while polypeptides offer the richest repertoire of self-assembly motifs.

However, irrespective of chemistry or assembly class, the synthesis of a discrete system that would span nano- to microscale dimensions is never a trivial task. Monodispersity, an ability to maintain the internal order and morphology of resulting assemblies and reliability of prescribed assembly modes remain the hurdles to be overcome towards functionally reproducible nanostructures.

Naturally occurring systems appear to be free of such barriers. They are highly conserved sequential couplings of exquisitely fitted subunits that use spatially self-maintained molecular arrangements. Emulating natural self-assembly patterns should be beneficial for engineering bioinspired systems. Nanobjects generated in this way can lead to materials with predictable and tuneable properties that are frequently referred to as "smart" materials. However, this hardly proves to be the case, in particular for *de novo* nanodesigns, which, despite their impressive numbers, remain short of original examples.

Indeed, whereas the total number of particular designs may have exceeded hundreds, unique designed forms are limited to just a few. In part, this is determined by applications, but possibly to a larger extent by the synthetic inaccessibility of large biomolecular subunits of natural assemblies. Thus, the success of extracting one robust rationale for artificial designs continues to be hampered by the need to find efficient ways which would allow for control over the assembly of smaller, simpler, albeit more entropy-dependent, self-assembling motifs. Given Nature's preference for biopolymer precursors, a set of requirements can be identified for a potential self-assembling candidate as follows.

First, it must be synthetically accessible in a monodisperse form. This requirement is limiting for any type of intended nanostructures, and also directly relates to the autonomous control of the nanoscale assembly.

Second, it has to adopt a recognition pattern ensuring a minimized impact of entropy factors (*e.g.* inter- and intramolecular dynamics) on the assembly, which ensures the hierarchical order of the assembly and consequently presents a major morphology-defining parameter.

Third, its assembly should obey the chosen mode of a hierarchical order encoded in primary sequences. This requirement is intrinsic to biopolymers but can be waived for certain molecular mimetics that preferentially lean on bulk forces supporting self-assembly, *e.g.* the hydrophobic effect.

There are several biomolecular motifs that can meet such design criteria. With their encoding traits established empirically, all manifest strong correlations between the chemistry and assembly. However, of notable advantage are those represented by two main classes: nucleic acids and polypeptides. Other motifs developed and used over the course of recent decades can be seen as derivatives or supplements of these two. Exemplified by just these two classes, the key factors underlying the functions of native nanostructures, including monodispersity, consensus folding and environmental responsiveness, provide impact on the choice of artificial designs. In conjunction with the growing body of synthetic designs and improving analytical techniques, the motifs stimulate the search for compatibility marks between structural principles of native assemblies and desired synthetic nanostructures. An attempt to address this or at least to touch upon some of the most design-responsive points in the prescriptive self-assembly is made in this volume.

The objective of this book is to outline developing approaches in the design of nanostructures, which employ bioinspired self-assembling motifs. Entitled *Bioinspired Nanoscale Design* or *Bionanodesign*, the book is written in the shape of a review referenced as fully as permissible within the context of biomolecular design, which forms a general trend throughout.

The volume is composed of three core chapters focusing on three predominant topics of applied nanotechnology where the role of nanodesign is prominent. Specific applications that arise from designed nanoscale assemblies as well as fabrication and characterization techniques are of lesser weight and whenever they appear serve as progress highlights for new functions.

The book places emphasis on "nano" rather than on "technology", providing recent examples of functions exhibited by nanodesigns, but which are not typical for naturally occurring analogues. This task is significantly eased by brilliant bioinspired designs reported to date and is complicated by the large volume of information created almost on a weekly basis. Therefore, the author apologises for the inevitable, but not deliberate, omission of examples many of which may prove to be equally if not more influential in bionanodesign.

References

1. G. E. Moore, Cramming more components onto integrated circuits, *Electronics Magazine*, 1965, **38**, 114.
2. G. M. Whitesides, J. P. Mathias and C. T. Seto, Molecular self-assembly and nanochemistry: a chemical strategy for the synthesis of nanostructures, *Science*, 1991, **254**, 1312–1319.
3. E. De Santis, H. Alkassem, B. Lamarre, N. Faruqui, A. Bella, J. E. Noble, N. Micale, S. Ray, J. R. Burns, A. R. Yon, B. W. Hoogenboom and M. G. Ryadnov, Antimicrobial peptide capsids of *de novo* design, *Nat. Commun.*, 2017, **8**, 2263.
4. G. Grossi, M. Dalgaard Ebbesen Jepsen, J. Kjems and E. S. Andersen, Control of enzyme reactions by a reconfigurable DNA nanovault, *Nat. Commun.*, 2017, **8**, 992.
5. R. P. Feynman, There's plenty of room at the bottom, *Eng. Sci.*, 1960, **23**, 22–36.
6. J.-M. Lehn, Supramolecular Chemistry – Scope and Perspectives: Molecules – Supermolecules – Molecular Devices (Nobel Lecture), *Angew. Chem., Int. Ed.*, 1988, **27**, 89–112.

CHAPTER 2

DNA as a Self-programming Module for Nanodesigns

2.1 Introduction

Five research papers published within a span of 3 months[1–5] defined the way in which biology is taught today. The importance of solving the molecular structure of nucleic acids has been discussed in numerous reviews and books. There is not a biologically relevant discipline which does not benefit from the knowledge of the DNA structure. However, none is probably more dependent on the accuracy with which the geometry and spatial organization of DNA are predicted and described than bioinspired nanotechnology. Over a few decades, DNA has emerged as a leading tool in nanodesign.[6–11] This chapter gives an overview of design principles and relevant examples using DNA, starting from design topologies,[8] algorithmic self-assembly[12] and DNA origami.[13]

DNA can be viewed as a repository of the information that genes carry and require passing to offspring. Invariably, this function determines the choice of the structural parameters and folding paths of the molecule. These are conserved and need to accommodate a faithfully reproducible mode of self-replication which can be translated into the building material of life, *i.e.* proteins. Rules to replicate life are expressed in the genetic code *via* the assignment of a unique codon, a triplet of nucleotides, to each of the 20 proteinogenic amino acids.[14] Strictly, there is more than one genetic code[15,16] and also more than one mode of DNA base paring.[17] However, those are particular cases and have yet to find evidence of impact within the context of DNA nanotechnology. More important in this regard are the facts that (1) only a part of genetic information is encoded by the code and (2) each human

Nanoscience & Nanotechnology Series No. 53
Bionanodesign: Old Forms for New Functions, 2nd Edition
By Maxim Ryadnov

Published by the Royal Society of Chemistry, www.rsc.org

cell type (apart from stem cells) expresses only one set of genes despite having the full copy of the genome. Furthermore, the genome is believed to contain the so-called "pseudogenes",[18] inactive and non-expressible parts of the genome that are often thought of as an evolutionary artefact or "junk" with no functional purpose.[19] The term is debatable as "non-coding" DNA accounts for about 90% of the human genome and can be a stored material with an unidentified function or a result of competing chromosomes.[20–22] Either way, this may prove to be important for nanodesign as the functional uncertainty of junk DNA as opposed to translated DNA can relate to structural alleviations observed for non-coding DNA structures, that is, the requirement for protein-coding DNA, which is read from one end, to be a linear molecule can be waived for non-coding DNA.

This implies that DNA architecture is intrinsically amenable to different topologies and shapes, the repertoire of which is inexhaustible. Whether natural or artificial, DNA designs consider the detailed understanding of DNA chemistry in addition to the visionary acceptance of the structural hierarchy of DNA.

2.2 Chemistry of DNA

Functional DNA is a monodisperse polymer composed of three types of repeating units – carbohydrate (deoxyribose, pentose monosaccharide); a heterocyclic base, which can be one of four: adenine (A), cytosine (C), guanine (G) or thymine (T); and phosphate – which together make up one DNA monomer, a nucleotide, Figure 2.1. Therefore, polynucleotide is an alternative name for DNA, commonly used as its chemical definition. The sequence of phosphates and carbohydrates coupled alternately constitutes a polynucleotide backbone which is decorated with heterocyclic bases or just basis linked to the first carbon atoms of the five-membered pentose rings, Figure 2.1.

Importantly, a phosphodiester bond lying between two carbohydrates is asymmetric as the bond links different (third and fifth) carbon atoms of the two. This renders a polynucleotide directional and confers the signature of DNA – the memory of chain direction. For example, an individual polynucleotide chain is a single-stranded DNA (ssDNA) that can form relatively flexible structures. However, these are thermodynamically unstable, and tend to intertwine with other ssDNA structures or themselves. A classical double-helix DNA (double-stranded DNA, dsDNA) is an antiparallel assembly of two independent polynucleotide strands oriented opposite to one another. The shape and stability of the structure depend on the extent of inter-strand interactions, which are provided by hydrogen bonding between the bases of the opposite strands, normally referred to as base pairings. The complementary base pairing as postulated by Watson and Crick[3] involves highly specific A–T and G–C interactions, which have proved to be sufficient for programming a large and diverse set of polynucleotide sequences confirming the robustness of this type of binding. The donor–acceptor patterns of hydrogen bonds are

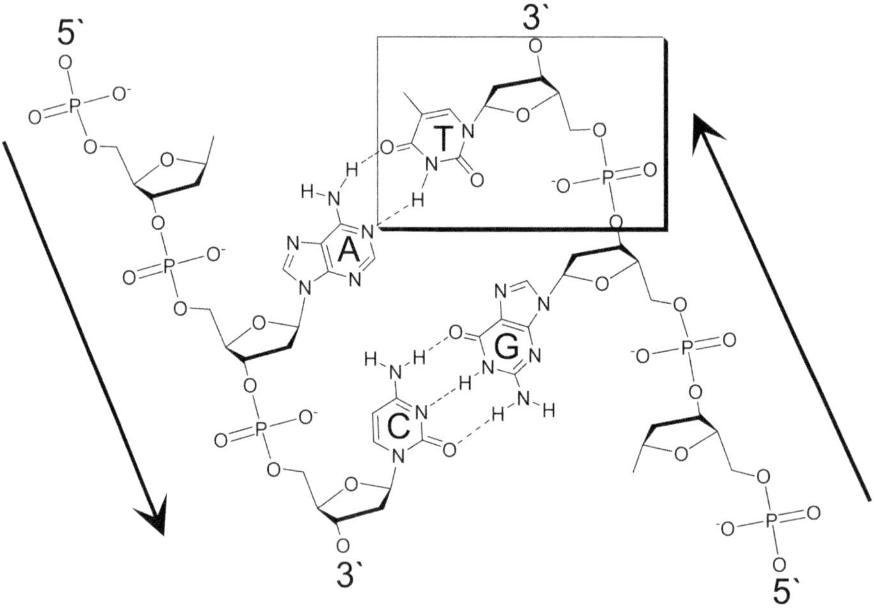

Figure 2.1 Chemical structure of DNA. Two polynucleotide chains run antiparallel to form one double-stranded DNA *via* Watson–Crick base pairing, A–T and C–G. For simplicity, only two pairs are shown. Dashed lines indicate hydrogen bonds. The square highlights a DNA monomer – nucleotide monophosphate, consisting of a base (T), a five-carbon pentose ring, deoxyribose and a phosphate group linking two pentoses. Arrows point in the $5' \rightarrow 3'$ direction of the asymmetric ends.

not identical for the two base pairs and differ in geometry and the number of bonds per pair. The A–T pair is formed by two bonds, whereas the more stable G–C uses three bonds. G and A belong to a heterocyclic family of purines, which are larger molecules than pyrimidine derivatives C and T. Owing to the size and the geometry of binding, purines can only marry pyrimidines and *vice versa*. These two parameters sum up a simple mechanism regulating appropriate pairings along polynucleotide sequences, Figure 2.1.

The conformational preference of the backbone of dsDNA is that of a polymer chain with conformational freedom considerably restricted by the regular stacking of sugar moieties. dsDNA folds as a right-handed helix with two distinctive grooves, major and minor, $\sim$2.2 and $\sim$1.2 nm wide, respectively, Figure 2.2. This so-called B conformation is the most common of three, which also include A,[5] a more compact or dehydrated form of B, and Z,[23,24] a transient left-handed zigzag structure. These structural criteria favour a predominant dsDNA conformation, the antiparallel B-form, Figures 2.1 and 2.2. An ingenious consequence of this form is that it endows dsDNA with remarkable stability and reconstruction properties, allowing it to function in cellular environments.

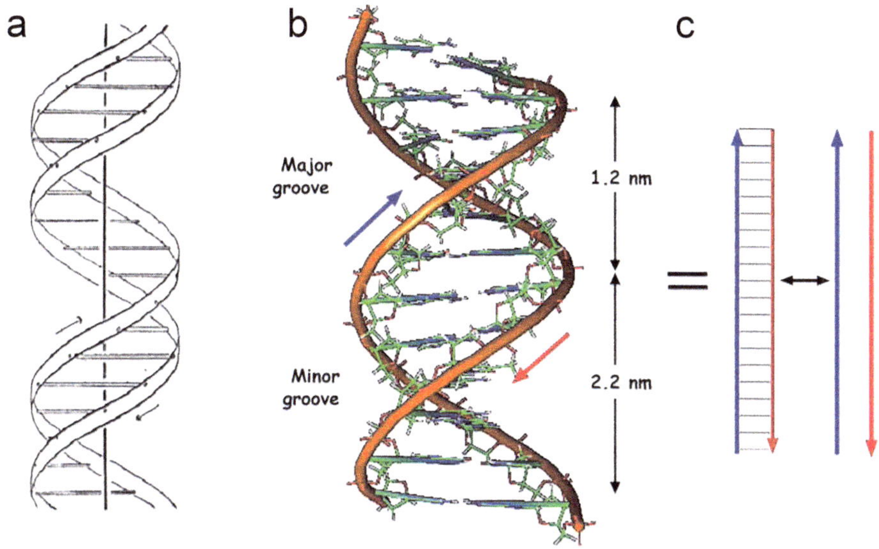

Figure 2.2 DNA double helix. (a) As originally proposed by Watson and Crick.[1] Reproduced from ref. 1 with permission from Springer Nature, Copyright 1953. (b) B-form of DNA as idealized from the PDB. (c) 2D ladder-like structures with arrows denoting 5′ → 3′ strands, with base pairs depicted as ladder steps (left) and, as equally used, omitted (right).

This is indeed a very stable molecule, but its stability carries a cost of maintaining complementarity, a constraint, which ensures the reconstruction of dsDNA from one of its strands. Rendering DNA materials self-replicating and self-repairable significantly increases the probability of error-free designs, which is consecutively supported by another strikingly simple mechanism. The binding energy of two strands can be approximated to k_BT per base pair, where k_B is Boltzmann's constant and T is a variable, *i.e.* temperature.[25–27] Given that under physiological conditions thermal changes exceeding $30k_BT$ are not common, double helices with more than 30 base pairs resist temperature-induced separations.[25,26] Restrictive endonucleases, or restrictases, target DNA, with each restrictase having a specific DNA cleavage site. Therefore, by cleaving DNA the enzymes reproducibly generate fragments of known sizes and sequences irrespective of DNA origin. This property offers tremendous potential for programming single-molecule constructions across length scales, which is one of the main reasons for the popularity and success of DNA nanotechnology. The property is inherent to DNA and remains true for any topology into which the double helix can be shaped. DNA is adaptable to virtually any supramolecular topologies and architectures provided that these are applicable to lengths exceeding the persistence length of the helix, estimated to be 150 base pairs or 50 nm.[28,29] Below this size, DNA is considered to be rod-rigid and inflexible, *i.e.* straight. However, owing to complementarity and the considerable flexibility of ssDNA, a common practice is to assign a given assembly mode to one strand, which complements the other strand in building up a desired morphology. This can vary depending on whether the desired assembly is a spatially fixed scaffold or a conformationally transient device, which is a matter of choosing a particular DNA motif.

Base sequences also play an important part in contributing to the flexibility of DNA.[30] G–C-rich regions give stiffer helices than those made of A–T pairs. The affinity of complementary sequences can be increased owing to the higher contents of G–C pairs, making a desired assembly more stable. However, this is not obvious and depends on correlations in hybridization between desired and unwanted conformations. Hybridization is a thermodynamic process of pairing two strands into a double helix. If the strands are complementary, their hybridization is energetically favourable. The event of folding two strands into a stable duplex is normally referred to as annealing to reflect its thermodynamic nature. Perfectly complementary strands anneal cooperatively under benign conditions. Their disassembly can be induced by thermal separation, but this would require transition temperatures, also known as melting points, which are substantially higher than those needed to reverse the annealing of strands with lower complementarity. Therefore, in designing DNA constructions, the stabilities of intended and unwanted hybridizations should be correlated and made maximized and minimized, respectively. This can be done by restricting similarities between complementary sequences and within individual sequences containing regions that are not meant to anneal.[31] The latter is particularly important in designing topology-driven DNA constructions, which use non-linear motifs with higher requirements for complementarity. In this light, thermal bending is another determinant to consider for designing different DNA shapes.[32,33]

Motion caused by temperature is the force behind the subtle deformations of the DNA helix. These deformations are subtle, but by being constantly excited are sufficient to expose restriction sites,[32] which serves biology. However, structurally thermal bending can be accompanied by the generation of permanent bends.[33,34] These occur without distorting the helix, which is attributable to sugar–phosphate backbones behaving as fixed strings retaining helical shapes. The strings can be loosened only when fully stretched under an external force. Under normal conditions, thermal bending tends to fold helices that are substantially longer than the persistence length into random coil conformations. In this case, the contribution of unusual shapes to the helix rigidity is minimized. This can constitute a mechanism that makes the meeting of distant sequences of the same molecule potentially useful for the assembly of extended closed systems.[35]

In one way or another, the design of novel DNA structures and devices relies on DNA motifs or structure-defining phenomena such as thermal bending. The coiling of two complementary strands into a rope-like double helix is the native geometry of DNA, but not the only one.

Inspirations for other geometries come from a repertoire of DNA assembly motifs found in both functional and in particular "junk" DNA, whose lack of biological function can make it a principal source of geometrical and topological templates. Nonetheless, even if restricted to the motifs of functional DNA, the number of design principles is more than guiding. With the development of high-resolution microscopy techniques such as atomic force, transmission electron and cryo-electron microscopy, it did not take long for artificial designs to capitalise fully on the virtually unlimited diversity of DNA topological motifs. Intermediate structural elements such as supercoils and knots, splits and junctions, crossovers and catenanes implicated in different DNA processes give the basis of the concept of a topology-driven DNA assembly.[8,9,31,36–39]

2.3 DNA Topology Motifs

With a variety of DNA topologies identified to support various biological functions, the representation of DNA as a double helix proves to have become a simplification.[6,7] In part, this is true when non-linear motifs are considered or when their relatively advanced complexity is taken into account.[39,40] However, all known and probably future motifs are variations of the double helix, which remains the main building block of the DNA architecture. Double-helical regions form the bulk of most DNA structures, which differ only in the number and type of topology elements. The careful choice of base sequences makes it possible to rationally design such elements or replicate their natural predecessors. For example, sticky ends – single-stranded binding overhangs of a duplex – which represent a common instrument in genetic processes such as transcription and recombination, are a common tool for DNA assembly and recombinant technology.

Similarly, sticky-ended designs appear to be a prerequisite for nanoarchitecture designs.[6] Other motifs involved in natural DNA processes are also common models in nanodesigns, including DNA junctions, loops and hairpins.[8,9] Building nanostructures based on these "primary" motifs relies on the reciprocal exchange of strands multiplied into crisscrossed or branched domains, all of which are realized through duplex formation.[37] These can be used to generate individually discrete architectures including catenated polyhedra,[41] single- and double-stranded knots[38] and Borromean rings.[42,43] Higher order forms or "secondary" motifs such as crossover tiles,[44–46] edge-sharing[47] and paranemic cohesion[48] belong to the next hierarchical level of architectures with periodic morphological features that otherwise may not be acquired.

However, irrespective of their complexity, DNA nanodesigns start with the choice of base sequences. One particular factor, which remains a priority in designing discrete nanoscale structures, that ensures high yields and monodispersity of assembled objects, is a prescribed hybridization. Correlations between melting points of target and undesired constructions present and assembly kinetics are other factors to consider. The folding of ssDNA, which for longer sequences becomes more autonomous and less predictable, prompts a systematic analysis or qualitatively more radical approaches.

The latter are favourable in sequence selection, with several prediction programs having been developed. These operate on either maximum-dissimilarity algorithms[31,49] or deduction approaches based on thermodynamic stabilities of designed folds.[50–52]

As an exemplar, DNA word design strategies[53,54] make use of existing thermodynamic models reflecting stabilities of formed duplexes and probabilities of secondary structures. In these algorithms, each DNA "word" – typically one DNA molecule of 12–16 nucleotides – is a tandem unit concatenated with other "words" into longer sequences.[55] A key objective here is the generation of large libraries of individual DNA constructs by incorporating multiple "word" variants at each tandem position. The format is thus combinatorial and allows the assembly of individual non-interacting strands assembled from individual "words". Specific strands of the libraries can then be selected based on their predicted and experimentally verified hybridization behaviours. In a similar manner, SEQUIN – a sequence-symmetry minimization algorithm[31] – selects sequences as constructed from subsequences of given lengths. The caveat here is to make each individual subsequence available only once in all possible sequences with no occurrence of the complementary subsequence. To achieve this, the subsequences are prescribed as overlapping and very short, typically 3–5-mers. The procedure is stepwise and implies the assembly of a target sequence from a collection of subsequences remaining in each step. This program presents a text-line interface, with sequence optimization being only semi-automated and requiring the involvement of the user throughout the process.

Other popular and interrelated examples include DNADesign and SpuriousDesign,[56,57] which are of elementary interfaces to the Vienna RNA package,[58] the core of which is a set of routines for identifying most optimized RNA secondary structures. The routines are accessed through C code libraries linked to independent programs (*e.g.* RNAfold, RNAdistance).[58] In DNADesign, MATLAB-based algorithms are utilized for the analysis of complementary subsequences optimized with user-defined scoring functions. SpuriousDesign, or rather its compiled C program SpuriousC, offers a faster interface and typically uses negative design principles that are primarily focused on minimizing "spurious" or unwanted interactions of undesirably complementary subsequences. The codes are efficient for designing relatively simple DNA folds and can be used for designing secondary elements for more complex and extended assemblies. However, as for all procedures of the type, each particular case normally requires the development of specific routines, the direct implementation of which on the interface may not be straightforward.

This can be improved by applying the codes into a graphical user interface, which ideally should lead to a fully automated sequence optimization. TileSoft[59] is a package featuring an evolutionary algorithm, which chooses "the fittest DNA sequences" from randomly generated populations. These individual sequences are "mutated" to give new sequences that are reinserted into the same population. The process is repeated, that is, evolves, until a termination point is reached. TileSoft may be viewed as optimized SEQUIN and DNADesign – same-sequence minimization procedures that are exponentially weighted within a library through random to fittest modifications – but its prime advantage is the integration of the user interface with its optimization module, which allows direct visualization of optimized sequences. Therefore, this algorithm is amendable to designing more complex structures, which can also be done in parallel and in multiplets.

Multimerization methods[60,61] offer somewhat higher potential for designing non-linear DNA motifs. These methods use an assumption that short DNA fragments can be multimerized into different topologies of various fixed lengths, and start with finding an equilibrium in distributions of topological states of a non-linear motif, namely circular DNA.[60] Circular DNA, which is described later, is a common natural topology and can be readily reproduced synthetically by ligation of linear fragments. Cyclization primarily depends on the length and bending of DNA.[62,63] The length can be defined empirically, whereas bending can be intrinsic or induced chemically or through binding interactions with protein ligands or other substrata.[63] Therefore, apart from deriving estimates for standard structural parameters such as the persistence length, helicity and complementarity, an additional parameter, fragment bending or the bend angle, is used to establish the distribution of multimers.[35,62] The bending of ligated fragments in this case is evaluated as a result of introducing specific irregular structures, which allows for a quantitative analysis of resultant structures and their structural features following the ligation of fragments into linear and circular multimers. The efficiency of cyclization is assessed to establish the conformational preferences of the structural elements within the generated multimers. The cyclization efficiency is specified by the ring closure probability, which is referred to as a *j*-factor – an experimental value that can be directly measured for each individual multimer.[64,65]

The *j*-factor is defined as the ratio of two equilibrium constants:[66] one for cyclization, K_1, and one for biomolecular association *via* two complementary sticky ends, K_2:

$$j = K_1/K_2$$

The *j*-factor equals the effective concentration of one end of a linear chain in the vicinity of the other end in the set angular and torsional orientations.[66] Reactions that join the two ends are faster than the covalent closure of the sticky ends. Therefore, the value of a given *j*-factor can be computed based on the conformational parameters of an individual DNA fragment, which can be obtained experimentally and related to measured properties of DNA fragments. The efficiency of such a methodology can be demonstrated not only for well-characterized linear duplexes[61] but also for more complex non-linear topologies (*e.g.* double crossovers).[67]

There are other algorithms similar to those described, and more can be expected to emerge, particularly given improving automation.[68,69] Most of the programs are efficient enough to convert the design process into interactive, automated procedures. However, the output capacity of virtually all of them is limited to the sequence level, which can be exemplified by the principle of DNA computing.[70]

The principle aims to solve the Hamiltonian path problem using traditional molecular biology protocols, and postulates that a set of rules can program DNA hybridization reactions into self-assembling systems.[70] Thus, DNA encodes a directed graph (digraph) in which different DNA constructs would give a set of all possible paths. The idea is to connect, *i.e.* assemble, the paths into a single combinatorial step, *i.e.* hybridization, and thereby different constructs would constitute different vertices and edges of the digraph, giving rise to a traceable sequence of vertices. Such a directed propagation occurs spontaneously, as self-assembly does, resulting from hybridization reactions, with each giving a DNA duplex representing a valid path in the digraph. When connected, DNA gives mathematically predictable patterns. However, the method assumes canonical DNA polymerization and hence is a generalization of the linear DNA assembly, which limits the model to finite-state automation.[69] As a result, the outlined programs are not deemed universal to support more sophisticated levels as additional and more complex parameters such as spatial orientation and topological directionality remain to be prescribed or correlated intuitively and empirically.

In this light, the importance of understanding how DNA structural motifs impose additional requirements becomes more apparent. For example, DNA can randomly form hairpins, which can be stable enough to inhibit hybridization, in particular for the shortest tri- and tetranucleotide loops. This effect is common in DNA assembly but appears to be largely neglected in algorithm developments, and so are internal and bulge loops.[69] Given that the formation of internal loops can be accompanied by the stepwise stacking of non-canonical base pairs, this should not come as a surprise.[71] Stacking interactions can also be symmetry driven, facilitated by helical twist angles, which impacts on the stability of loops of different symmetries but the same lengths. DNA bulges – junction loops that can be as short as one nucleotide – have been studied more systematically, but their structural parameters can only be approximated using a series of assumptions.[72–74]

These are just a few examples demonstrating known DNA motifs, which are to be implemented in prediction algorithms. Different topological forms of DNA are by no means contingent, but are essential components of biological systems.[6,22] These are non-linear elements and are spatially constrained molecules. Linear duplexes may appear as less constrained but are adjustable for topology control at the secondary structure level, which eventually determines the rich variety of unusual DNA shapes. Consequently, constructing larger objects on the nanoscale compiled from geometrically predefined units requires a coherent means of self-assembly and covalent conjugation, with emphasis on DNA self-assembling properties.[8,9] Therefore, DNA motifs can be liberally separated into two categories. For the purpose of this chapter, these are given as 2D orthogonal variations of linear DNA, "primary" motifs, and their posterior assembly forms or "secondary" motifs that are directly used as building blocks in the construction of discrete nanosystems.[6]

2.3.1 Primary DNA Motifs

Before describing primary motifs, it is important to restate the role of sticky ends in DNA assembly, Figure 2.3a. Sticky ends constitute the simplest primary element with a universally important function for DNA assembly.

First, sticky ends offer the most reliable set of intermolecular interactions available in self-associating systems, while complementarity and affinity principles are straightforward to translate into specific sequences. The sequence designs will maintain the local structure of the antiparallel B-DNA, a building block with structural parameters tuned with sub-nanometre precision.

Second, the length of sticky ends can vary from a few nucleotides to tens to more than 100 residues without compromising their affinity and binding properties. The range of routine synthetic and enzymatic techniques is available to construct desired overhangs with a selectivity of a single residue.

Finally, sticky ends are compatible with other primary elements of practically any geometry, symmetry, structural regularity and size, which makes the sticky-end motif a versatile precursor for all nanoscale design forms.

The next simplest and remarkably impactful primary motifs for DNA nanodesign are axial DNA topologies – junctions and circular DNA, Figure 2.3.

As discussed earlier, the double helix is a linear molecule, that is, it is axially straight. However, DNA junctions can be readily generated. One of the most common examples is three- and four-arm junctions whose architectures rest on three and four complementary ssDNA, respectively.[8] The arms of these structures, ephemeral in DNA replication and recombination,[75–77]

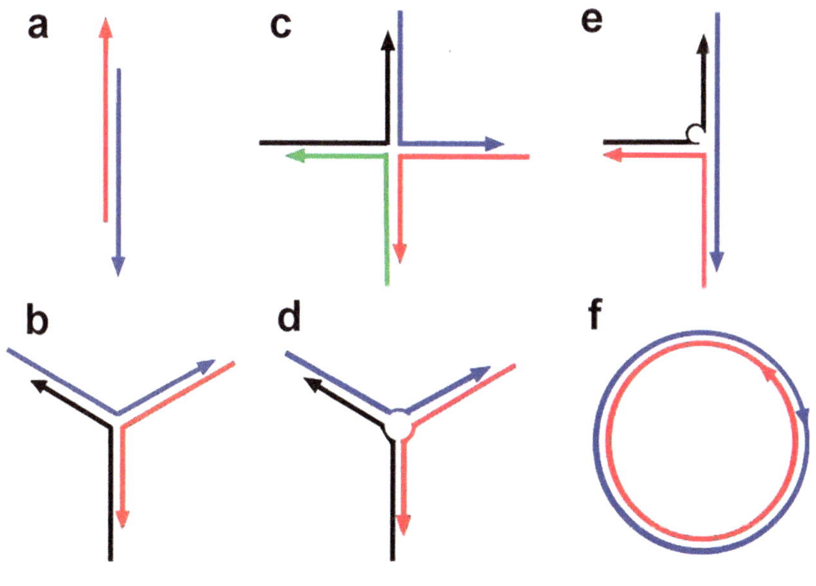

Figure 2.3 Primary DNA motifs. (a) Sticky ends. (b)–(e) Junctions with (b) three arms, (c) four arms, (d) internal loop and (e) bulge loop. (f) Circular DNA.

are dsDNA, with each being made of halves of two different ssDNA. Thus, one ssDNA participates in two different arms, Figure 2.3b and c. This allows connection of the helices by sharing one branching point. Higher generation junctions with a greater number of arms can also be made, but these tend to be less stable and structurally unpredictable.[78,79] Extended assembly of immobile junctions that lack homologous sequence symmetry is used in the fabrication of more complex networks and lattices.[8,9] This is done by terminating the arms in sticky ends to promote their self-assembly and following ligation, which makes the procedure identical with that used in standard linear DNA, Figure 2.3. All such motifs are characterized by the degree of their "connectedness", defined as the number of edges, *i.e.* double-stranded arms, which merge at a vertex, *i.e.* branching point.[8,78,79] The number of arms used in a junction is what limits the connectedness of the vertex.

Another important feature is that the mobility of a given vertex can be controlled by the incorporation of smaller motifs prescribed in the sequence. This can also determine the mechanical properties of a junction. For example, the inclusion of unpaired bases in the strands at vertices would yield internal loops resulting in a less stiff junction, Figure 2.3d, whereas a bulge loop introduced in just one strand can fix the angle between two arms to 180° induced by edge-stacking interactions, Figure 2.3e.[80,81]

Branched junctions of the described types have been used as building blocks for a number of *de novo* DNA topologies, including a quadrilateral,[82] a cube[83] and a trefoil knot.[84] These are catenated and knotted motifs as described below.

Success in constructing these objects inspired other designs based on multistranded and double-helical DNA in which each strand contributes to two different helices.[85] This led to the so-called anti- and mesojunctions,[86,87] which represent generic models of naturally occurring multistranded DNA complexes such as the Holliday recombination intermediate.[77,86,88] Unlike in a conventional junction, only some and none of the axes in a mesojunction and an antijunction, respectively, point towards the centre of the junction. One consequence of this innate feature of the complexes is that geometric populations have different numbers for different values of maximum connectedness, which increase as the number of arms increases. This makes designs based on the types difficult to predict and difficult to assign to specific architectures, which bears similarities with crossings in knot designs.[38,89] Conventional junctions may appear disadvantageous in the construction of regular grid-like nanoscale constructions such as lattices, matrices and dendrimers.[90,91] Structural irregularities typically observed in such cases can be attributed to the conformational instability of branching points, impacting on the physicomechanical properties of junctions.[90,91]

Another seemingly simple but important primary element is circular DNA, Figure 2.3f. This is the element of choice for bacterial DNA molecules,[92] that is, plasmids, which are used as gene vectors in genetic engineering. It also is the form of mitochondrial DNA of eukaryotes and is a usual constituent of macro or "giant" DNA, protein-fused multiloop structures.[93] Biological functions of circular DNA are diverse, but structurally two forms of circular DNA are recongnized,[94] which are designated forms I and II. The former is a closed form, that is, each of the strands is closed on itself. Form II is a less compact derivative of form I occurring as a result of a single-stranded break.[95]

The formation of the closed form with both strands being intact requires a specific topological state. The cyclization of a linear chain entails the setting up of a conformational constraint.

In polymer physics, a cyclized polymer chain can be described as a knot.[96] With the conformationally fixed double helix, DNA knots are formed with respect to the helix axis. The double helix depicted in Figure 2.3f can therefore be viewed as a trivial knot. In contrast, the strands of "relaxed" DNA are interwoven with a standard twist around the axis which can be altered by a strain.[97] Cyclization itself provides a sufficient stress for the double helix to twist out of the circular form. This is achieved through supercoiling[98] – a process of global deformations of circular DNA that help DNA relax.[35] The deformations result in the helix being wound or rotated around its axis a number of times equivalent to the number of accommodated helical twists.[63] Such rotations are called writhes.[99] One writhe is the number of helix intersections or coils, whereas one twist is the number of helical turns. In this notation, supercoiling can be simplified as a sum of mutually convertible writhes and twists, which allows for the description of the cyclized double helix as a torus, that is, a circle rotated around its axis.[99,100] The number of rotations in this case can serve as a basis for the classification of the resultant knots.[101,102]

Since in a circular DNA both strands are linked, the link between them can give a quantitative parameter to model supercoiling or describe it mathematically.[33] This can be done using a "linking number", *Lk*,[99] which is the algebraic number of intersections between one strand and the edge of an imaginary surface created by the other strand.[100] Quantitatively, *Lk* is close to the sum of helical turns and writhes and depends only on the strand topology. Consequently, it remains unchanged through all conformational changes of a closed cycle and its value is always an integer.

These principles underpin the rationale of the knot theory adapted for DNA knots.[102,103] The approach uses a further assumption that an individual knot must be deformed to the standard form of its projection on a plane, whereas deforming a knot does not change it.[98] The standard form is an image created by a minimum number of crossings without self-intersections.[102] Thus, any knot can be defined according to the lowest number of intersections of any of its possible diagrams, its crossing number,[102] Figure 2.4. The simplest knot would have three intersections in its standard form and there would be only four types of knots with less than six crossings,[102] Figure 2.4.

The number of knot types increases dramatically with increase in the number of intersections. For example, there are 49 types of knots with 9 crossings, 165 with 10 and 552 with 11, which in total amounts to 1 701 936 knots with less than 17 crossings.[102,104] Knots are catalogued by their crossing numbers with a subscript stating a particular knot shown out of many

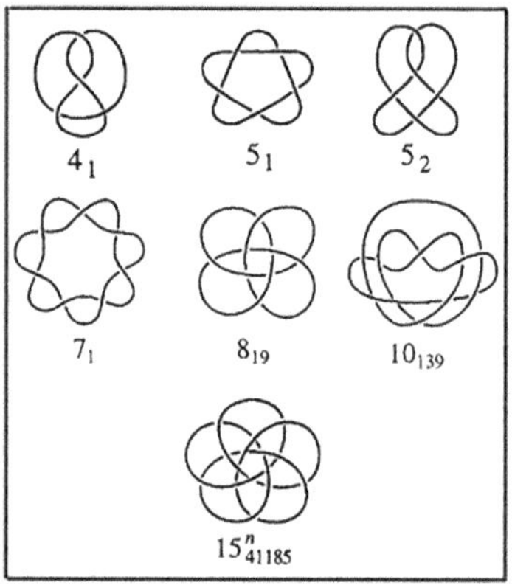

Figure 2.4 Topological states in circular DNA. Standard forms of knots presented with different numbers of crossings in plane projection. Reproduced from ref. 60 with permission from National Academy of Sciences.

with the same number of intersections, Figure 2.4. These are simple or prime knots. The theory is also used to tabulate composite knots, which are created by cutting and then reconnecting arcs on two prime knots without losing their orientations.[102] However, these may be of lesser relevance to DNA nanodesigns.

The interpretation of circular and supercoiled DNA as knots is not of purely mathematical interest. Linking experimentally proven DNA knots with topological methods may lead to exploitable design rules.[105] Adapting the knot theory for *de novo* DNA designs may follow a similar path to that used to reveal the enzymatic supercoiling of DNA[106,107] or the mechanistic understanding of DNA replication and recombination.[101] Furthermore, topological approaches prove to contribute directly to establishing subtle differences in the behaviour of enzymes from closely related classes.[108]

For example, the unknotting of DNA by topoisomerase II is accompanied by a stepwise decrease in writhes.[107,108] The decrease consists of two steps of transient breaks in DNA, with the enzyme first cutting both strands on one side of a crossing point and then re-ligating them on the reverse side.[107] By contrast, for type I topoisomerases, which untangle DNA by cutting a single strand to rotate the generated free ends, and only then re-annealing them, the decrease in crossing number consists of a single step.[109] Thus, both topoisomerase types change the linking number of DNA, but do it differently: type I changes by one, whereas type II changes by two.[110]

The true potential of circular DNA in nanodesign may not have been exploited to the deserved extent. Possibly, part of the problem is the topological properties of the element. These are difficult to reproduce in experimental designs. As a result, the design of knotted DNA motifs remains challenging and requires a sound understanding of both the mathematics and chemistry of DNA assembly.[10,38]

The emergence of novel motifs advocates that nanostructures are best characterized by their topology, *e.g.* branching, rather than their geometry. This particularly helps the cause of applying DNA nanotechnology to the construction of motifs defined by their linking, *i.e.* knots. The ability to create a synthetic version of a crossing or a node is met by the double helix, with the relationship between crossings and DNA illustrated on the trefoil knot depicted in Figure 2.5a and b with an arbitrary polarity.[89]

Each crossing of the knot is placed into a square to convert portions within the squares into diagonals. The diagonals are meant to divide the square into four regions, with two of these regions being between parallel and the other two between antiparallel strands. Because the strands of the double helix are antiparallel, pairing has to be designed to run over a half-turn segment, *i.e.* six base pairs, in the regions between antiparallel strands. The formation of target crossings can thus be specified by complementary sequences. With two mirror types of crossings existing (positive and negative), two mirror types of DNA are required, Figure 2.5c.

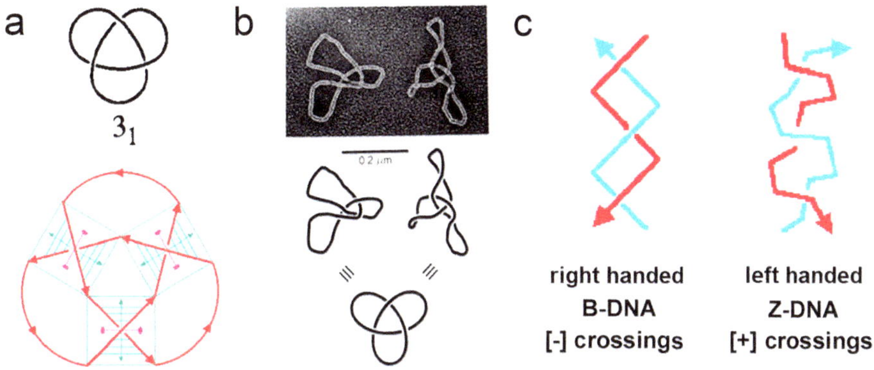

Figure 2.5 Design of single-stranded DNA knots. (a) Relationship between the crossings of a trefoil knot and B-DNA. Thick arrows indicate the path of the DNA. Each square shown in cyan lines is defined by a crossing between two arrows comprising four domains – two between parallel and two between antiparallel arrows. Antiparallel arrows are linked by six thin lines denoting six base pairs corresponding to a half-turn of an antiparallel B-DNA. Cyan double-headed arrows indicate helix axes. Reproduced from ref. 40 with permission from IOP Science, Copyright 1998. (b) Electron micrograph (top) and tracing schematics (middle) of trefoil knots and a redrawing of the knot with the minimum number of crossings (bottom). Each knot has three irreducible (−) nodes. Reproduced from ref. 106, https://doi.org/10.1016/S0021-9258(18)54963-2, under the terms of the CC BY 4.0 license https://creativecommons.org/licenses/by/4.0/. (c) Negative and positive crossings corresponding to the right-handed B-DNA shown as one and a half turns, and its mirror image: left-handed zigzag Z-DNA. Reproduced from ref. 40 with permission from IOP Science, Copyright 1998.

The main DNA form, B-DNA, is a right-handed helix and its crossings can be designated as negative. Although there is no DNA form known which is an exact geometrical mirror image of the B-form, topologically the left-handed Z-form can be used to generate positive crossings,[24,38] Figure 2.5c. The two DNA forms are qualitatively different and the B → Z transition does not occur spontaneously and must be promoted.[111] In natural systems, the lifetime of Z-DNA is very short but can be induced by specifying sequences and external conditions.[23,111] Although such conditions may not be benign and compatible with enzymatic ligation, it is possible to assign a Z-forming propensity to short segments by introducing CG repeats or modifying bases.[24] Using these parameters in a basic design framework, a rich repertoire of knotted DNA elements can be constructed.[6,85]

Two pairing domains containing one double-helix turn each can be merged into a single molecule. Both domains can switch to Z-DNA but one has a higher transition propensity and does it more readily than the other.[89] Different knotted species are thus generated by playing on the differences in the sensitivities of the domains towards assembly conditions, for instance, by changing the ionic strength in DNA solutions,[89] Figure 2.6.

Briefly, at very low concentrations of dications, none of the domains assembles into the double helix. At increased concentrations, both domains fold as B-DNA, giving rise to a trefoil knot with negative crossings. Further increases in concentrations create weak Z-inducing conditions under which only one of the domains, the more sensitive, converts into Z-DNA, generating a figure-of-eight knot. Strong Z-favouring conditions force both domains to adopt the Z-form, yielding the trefoil knot with positive crossings,[89,111] Figure 2.6.

An apparent drawback of this design is that all of the conversions are irreversible, and the formed topologies are final: changing the solution conditions cannot facilitate the transformation of one knot to another without cutting and re-ligating the covalent backbone.

This is precisely what type I topoisomerase acts upon. Therefore, characteristic stepwise interconversions of different knots, both from B- to Z- and from Z- to B-forms, could be performed.[111] The impressive success of this experiment prompted an attempt to fish out an RNA topoisomerase – a predicted but previously unfound enzyme – with the help of an RNA knot.[112]

Like DNA, RNA molecules can contain hairpins and pseudoknots.[113] Unlike DNA, cellular RNA is a single-stranded molecule. Therefore, it was thought that the formation of functional RNA structures was not hampered by the requirement to solve the RNA molecular topology problem, which does not exist in the cellular context. The search for an RNA topoisomerase did not seem necessary. However, since it was known that DNA topoisomerase III – a recombination-regulating topoisomerase – can cleave RNA, the use of the enzyme to catalyse the interconversion of RNA topological states is possible. To probe this, a cyclic RNA strand was designed to adopt two topologies – a circular RNA and a trefoil knot.[112] Because the two forms are the resultant forms of RNA folding, their interconversion can only be achieved by a strand passage event, which as for DNA knots can be catalysed enzymatically. Consistent

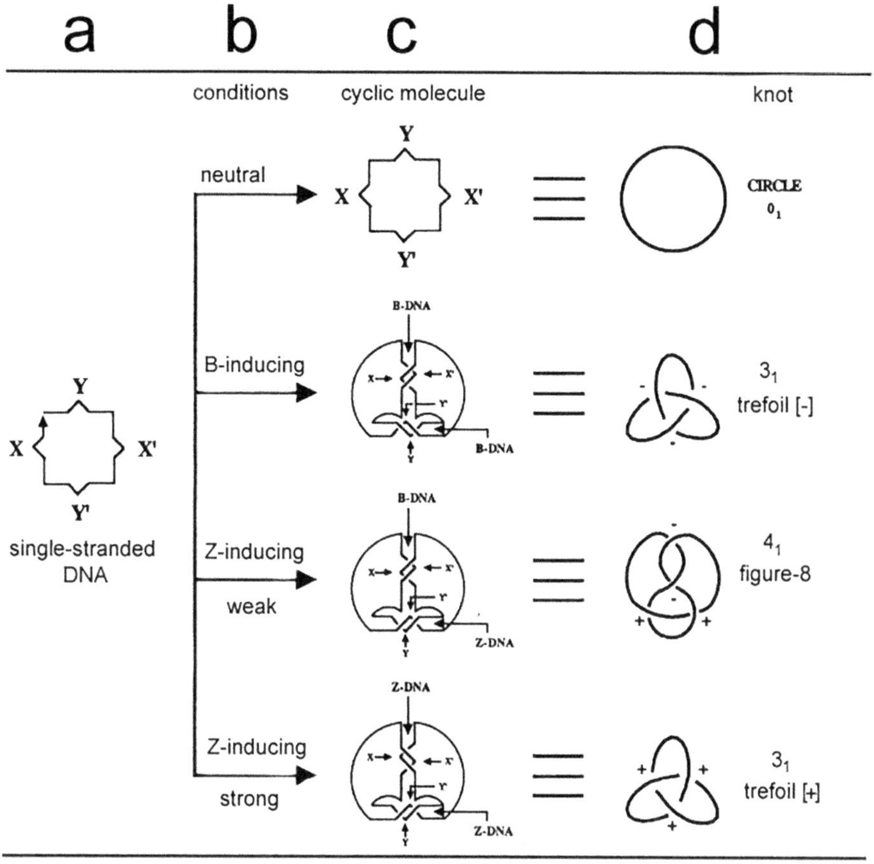

Figure 2.6 Folding of single-stranded DNA knots. (a) A single strand containing four regions, X and Y and their respective complements X′ and Y′; (b) four independent conditions of folding and ligation; (c) generated target products; (d) their corresponding molecular topologies. Adapted from ref. 89 with permission from American Chemical Society, Copyright 1995.

with this, the anticipated RNA strand-passage activity of DNA topoisomerase III accompanied by circle–knot interconversions was observed.

The role of topoisomerases in the topological expression of the double helix appears to be unequivocal and is evenly balanced between DNA unknotting and supercoiling. In fact, the enzymes support the formation of more complex structures that can be thought of as higher order forms of knots. These are catenanes.

Catenation is the process of interlocking of DNA rings catalysed by topoisomerases. Catenated states of DNA resulting from the compaction of DNA substrates and favoured at high local concentrations can be rationally directed in a fashion similar to that for DNA knots.[114] Indeed, knots and catenanes are closely related and it is relatively straightforward to interconvert them by a simple operation on a crossing. One crossing may consist of four polar strands connected in pairs, with two before and two after the crossing. Knots and catenanes become interconverted every time, when the crossing is removed through the pairs exchanging outgoing partners. As a result, DNA, which is used to construct knots, should be equally effective in constructing catenanes. This proved to be the case in both natural and synthetic systems. Catenanes as simple as dimeric[115,116] are common in natural systems and can be found as intermediates in DNA replication or products of site-specific recombination systems.[117,118]

Recombination by integrative systems, *e.g.* bacteriophage λ, is found to provide convenient methods for the synthesis of simple linear catenanes used in biochemical studies.[116] However, most of such catenanes are substantially more flexible molecules than knots and are not particularly compelling elements in designing geometrically fixed nanoscale scaffolds.

The problem is similar to that of topological engineering based on single-stranded DNA, which was solved by "topological protection" techniques involving the temporary complexation of single strands with their complements.[119]

For DNA nanodesign, most notable examples are higher generation catenanes, the topology of which is translated into more complex and ordered forms such as DNA stick polyhedra: DNA cube and a truncated octahedron.[41,83] However, these structures may be described only on the topological level of linking as geometrically they are not characterized by well-defined coordinates. For this reason, the structures in Figure 2.7 are sufficiently schematic.

In reality, in branching points the angles between the double-helical edges are looser than they appear. Each edge contains an exact number of double-helical turns, which makes each of the faces correspond to one cyclic single strand. With the cube composed of six different cyclic strands and the octahedron of 14, the objects become hexa- and tetradecacatenanes, respectively, Figure 2.7a and b.

In close agreement with the design conventions of topology-driven DNA constructs, the synthesis of catenated DNA can serve as a proof of topology and *vice versa*. Indeed, linear catenanes can be targeted through cutting restriction sites incorporated into specific edges of the structures. This breaks the structures down to individual catenanes, which can be characterized according to their electrophoretic properties.[120]

Figure 2.7c illustrates an experiment exemplified by the synthesis of the cube. In the first step, a linear triple catenane serves as a starting material corresponding to left, front and right faces of the cube. Once the cube has been synthesized, it is cut on the two front edges in the catenane to give another linear triple catenane corresponding to the top, back and bottom faces of the cube; if the synthesis is unsuccessful, the starting catenane is destroyed.[83] For the truncated octahedron, the strands equating to the six square faces were confirmed first, followed by restricting the octacatenane matching the eight hexagonal faces down to the tetracatenane flanking each of the squares.[41]

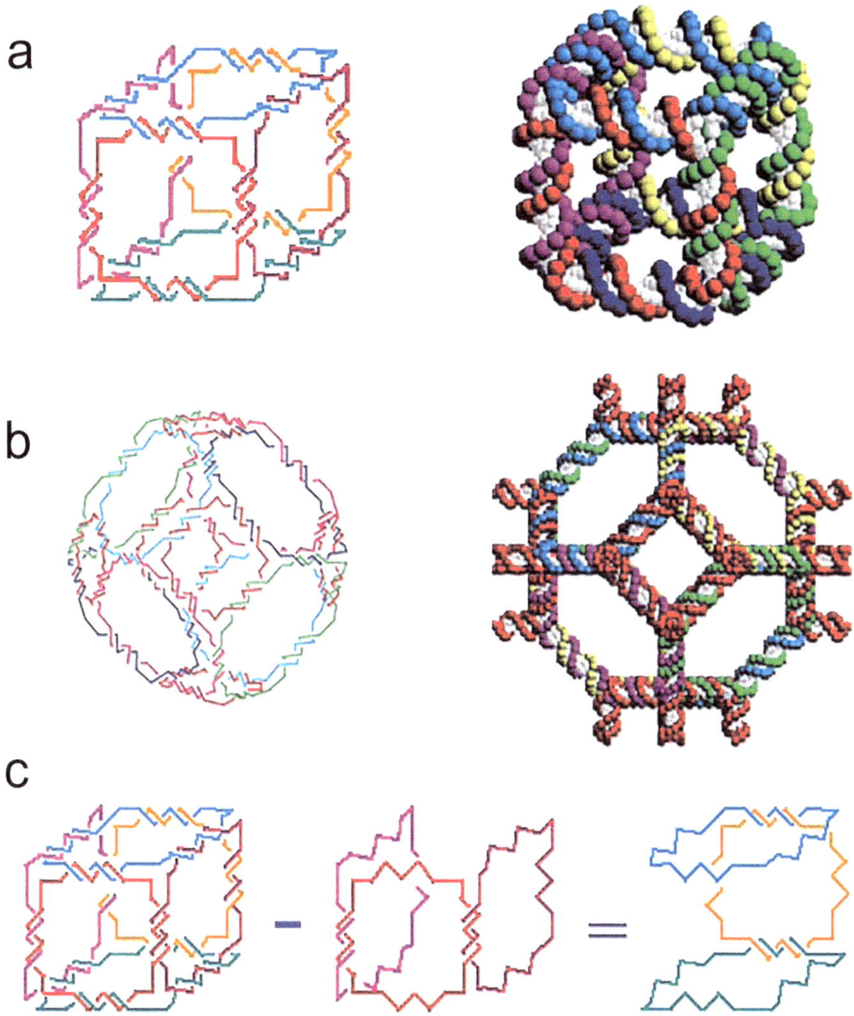

Figure 2.7 DNA catenane designs. (a), (b) Sketch (left) and model (right) representations of (a) hexacatenane with the connectivity of a cube and (b) tetradecacatenane with the connectivity of a truncated octahedron. (c) Formation of the cube through linking linear triple catenanes. Reproduced from ref. 39 with permission from Elsevier, Copyright 1999, and from ref. 40 with permission from IOP Science, Copyright 1998.

These examples demonstrate that catenanes and knots represent motifs, which are interesting *per se*, and some of them are versatile enough for designing nanoscale objects. However, their use in the construction of more difficult targets, such as secondary motifs and further static 2D arrays or 3D discrete materials, is restricted by the mechanically untranslatable geometrical "floppiness" of the generated crossings,[7,10,40] which at their present state makes them practically unsuitable for functional nanodesigns.

Nevertheless, in designing knot-related motifs, the DNA expression of one topology stands out as too challenging and promising to be omitted, namely the Borromean rings, a topological link named after the three-ring heraldic design of the Italian Renaissance Borromeo family,[43] Figure 2.8a. Considered as one of the most ambitious, the link presents a rich family of topological structures that are very similar to catenanes, although only geometrically.

The interbonding of rings in catenanes is analogous to that of the links of a chain. As in the chain, the breaking of one ring (link) separates the two parts of the chain flanking the broken ring. Borromean rings are topologically different. The hallmark of this motif is the inseparable arrangement of three interlocked circles or rings that can be unravelled into three independent pieces in one step with the removal of any of the individual circles. In other words, if one ring is broken the whole structure will fall apart, giving rise to a number of intact but unlinked circles. The critical nature of this key difference is evident from the synthetic point of view. Although the syntheses of catenanes and knots are more or less routine operations, the unique makeup of non-catenated, albeit interlocked, Borromean rings makes finding a universal synthetic strategy a formidable task.[121]

This may be a subject of future successful attempts and studies, but the sufficient and necessary criteria in tackling the problem start with establishing the direct relationship between the topology and its molecular adaptation for the chosen type of assembly. For DNA-based designs, the traditional representation of Borromean rings as planar Venn diagrams, in which the innermost three crossings are viewed as negative and the outermost three as positive, is modified, Figure 2.8.

For DNA, it can be redrawn to replace all of the crossings by one and a half turns of the double helix as opposed to a single half-turn in knot designs. The use of one and half turns is more convenient,[42] with each crossing replaced

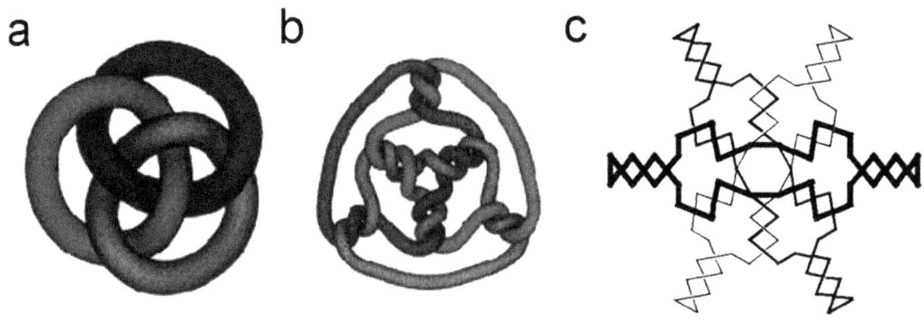

Figure 2.8 Design of DNA Borromean rings. (a) The traditional representation. (b) Borromean rings with each crossing replaced by three to reflect one and a half turns of the double helix. The inner helices are right-handed and correspond to B-DNA; the outer helices are left-handed and correspond to Z-DNA. Reproduced from ref. 40 with permission from IOP Science, Copyright 1998. (c) Schematic drawing of the Borromean rings assembled from two three-arm junctions. The scission of any of the three rings shown gives the unlinking of the other two rings. Reproduced from ref. 10 with permission from Springer Nature, Copyright 2007.

by three with the innermost segments corresponding to a three-arm branching point of B-DNA, Figure 2.8b. Therefore, the three inner helices are right-handed and the three outer helices are left-handed, which preserves the original equal number of positive and negative crossings.

Because any topological picture can be deformed an infinite number of times, the generated projection can be adjusted to give a clear retrosynthetic disconnection path involving the ligation of two stable three-arm junctions constructed from B-DNA and Z-DNA, Figure 2.8c. In such a structure, each of the rings contains a different hairpin motif. Arranging hairpins at the equator of the representation is convenient in assigning ligation and restriction sites of the two junctions, which can be of different sizes to facilitate the separation of restriction products, Figure 2.8c. As in both previous catenane-based designs, this feature ensures an unambiguous characterization of the resultant structure, providing the proof of the target topology. Specifically, when one ring is cleaved the other two remain but in a separate, *i.e.* non-bonded, form, which confirms the key feature of the Borromean rings.

The approach was tested experimentally in the synthesis of six ssDNAs corresponding to the three strands of each branched junction, B-DNA and Z-DNA, which were annealed individually and then ligated under Z-DNA-promoting conditions into the first DNA Borromean link.[42]

The method represents an excellent example of topological control that in principle may be incorporated into other biomolecular systems. Nevertheless, despite such an impressive execution of the topology using DNA, the motif shares the same design limitations that are characteristic of knotted and catenated structures. These motifs struggle to find use as generic building components for periodic structures featured over larger length scales.

Indeed, building more complex nanoscale objects demands additional geometrical constraints, which can support spatially better defined and molecularly more conserved cohesion patterns. The architectural frameworks of the primary motifs are almost exclusively mediated by sticky-ended associations. Generally, and as shown by the outlined examples, the sequence specificity of sticky ends proves to be a powerful tool in meeting a number of criteria for nanoscale construction, but these are set at local structural predictability, limiting the types of possible designs. Although this may be sufficient for engineering even infinitely long objects such as infinite double helices,[122] a larger scale design requires operationally more advanced motifs.

Elements that are complementary to the sticky-end cohesion and are capable of meeting requirements for periodic arrangements on the nanoscale are described as "secondary" motifs.

2.3.2 Secondary DNA Motifs

This term is introduced not only to reflect a higher order of secondary motifs compared with primary or individual elements, but also to stress the need for finding robust nanoscale components to match mechanically sustainable nanofabrication.[9] An archetypal representative of secondary motifs is the so-called double-crossover element.

This motif is a more complex topological expression of junction motifs. Interest in exploiting it for nanodesign is driven by the need to develop topologically tighter structures. The main efforts have been focused on devising different classes of junctions which have been explored fairly extensively. Elegant designs described earlier, anti- and mesojunctions[86] and also site-specifically constraint bulged junctions,[80,81] serve as classical examples. However, the discovery of a double-crossover motif, which has become one of the most impactful DNA elements, can be considered as the culmination point of the preceding attempts.[44,45]

This secondary element, abbreviated DX, is a recombinatorial motif related to the Holliday junction.[44,119] DX consists of two four-arm Holliday junctions joined by two double-helical arms at two adjacent arms, Figure 2.9.

There may be five distinct DX isomers, designated DPE, DPOW, DPON, DAE and DAO, Figure 2.9. The isomers differ by (i) having parallel (the first three) and antiparallel (the last two) orientations of their helix axes, and by

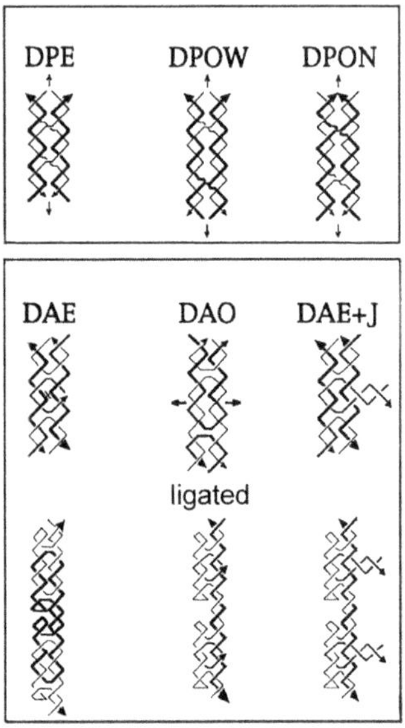

Figure 2.9 Double-crossover (DX) motifs. Parallel (top panel) and antiparallel (bottom panel) motifs are shown with zigzag helical structures denoting strands. Two consecutive perpendicular lines correspond to a full helical turn for a strand. Line thickness is related by symmetry elements shown as double-headed arrows – twofold axis lying in the plane of the page; and by 'x' (in the centre of DAE) – an approximate dyad perpendicular to the plane of the page, which for DAE+J is removed by the bulge. Thick lines in the ligation products of antiparallel motifs (bottom) correspond to reporter strands for DAE and DAE+J and catenated rings for DAO. Adapted from ref. 44 with permission from American Chemical Society, Copyright 1996, and from ref. 45 with permission from American Chemical Society, Copyright 1993.

(ii) the number of double-helical half-turns, odd (DAO, DPOW, DPON) or even (DPE, DAE), between their crossover points.[44] DPOW and DPON molecules also contain an extra half-turn corresponding to a major (wide) or minor (narrow) groove spacing. Thus, in the abbreviations given to each of the isomers, D, A, P, E, O, W and N stand for double, antiparallel, parallel, even, odd, wide and narrow, respectively.

Importantly, the DX molecules differ also in their behaviour, which helps define their preferential use in nanodesign.[44,45] The parallel double-crossover, transient motifs relevant to biological processes are not well behaved in non-denaturing gels,[44] particularly when the separation between their crossover points is short, unless their ends are furnished into hairpin loops.[123] The parallel orientation of the helical domains implies that the minor and major grooves of both engaged helices come together each turn, which leads to long-range repulsive interactions that destabilize the structures. Antiparallel DX, in which minor grooves of one helix are facing the major grooves of the other, give more stable structures.[45] Therefore, much research has been focused on antiparallel DX molecules – DAE and DAO.[6,22] The molecular topologies of these isomers differ only in the number of double-helical half-turns between crossovers, *i.e.* odd and even. These molecules are twice as rigid as linear duplexes[67] and can be used as oligomerizing elements for nanoscale constructions.[45] A third motif, DAE+J, introduced as the DAE motif containing one or more bulged junctions (+J) placed between the crossovers, Figure 2.9.[124] The bulge, which is composed of two nucleotides, enables an interrupted helix to be stacked and so to become an extra double-helical hairpin protruding from the plane of the two DX helices. This arrangement can be used as a topographic marker or reporter in microscopy and ligation experiments.

The application of this design principle was demonstrated in the design of DNA tiles.[124] DX motifs modified with hairpin junctions were used to tile a plane generating periodically striped 2D lattices, Figure 2.10.

One of the motifs used (DAE) with dimensions of 4 × 16 nm was shown to self-assemble into single-domain crystals of 2 × 8 μm with a uniform thickness of 1–2 nm. Because the vertical and horizontal repeats of the assembled arrays are 4 and 32 nm, respectively, and because features of 4 nm were difficult to capture at the time by atomic force microscopy (AFM), the protruding periodicities could only be observed at a distance of 32 nm, Figure 2.10b.

The helices derived from hairpin junctions and sticking out of the plane generated visible stripes above the surface at these intervals, Figure 2.10c. The pattern was highly periodic and additional modifications were revealed to give a larger spacing when using different components. Specifically, when four components (instead of two that gave the 32 nm spacing) were used, a spacing of 64 nm was obtained, Figure 2.10c.

The approach shows that it is possible to control surface topology and such self-assembling lattices can be used as scaffolding materials for templating the assembly of other systems through chemical or molecular recognition recruitment.

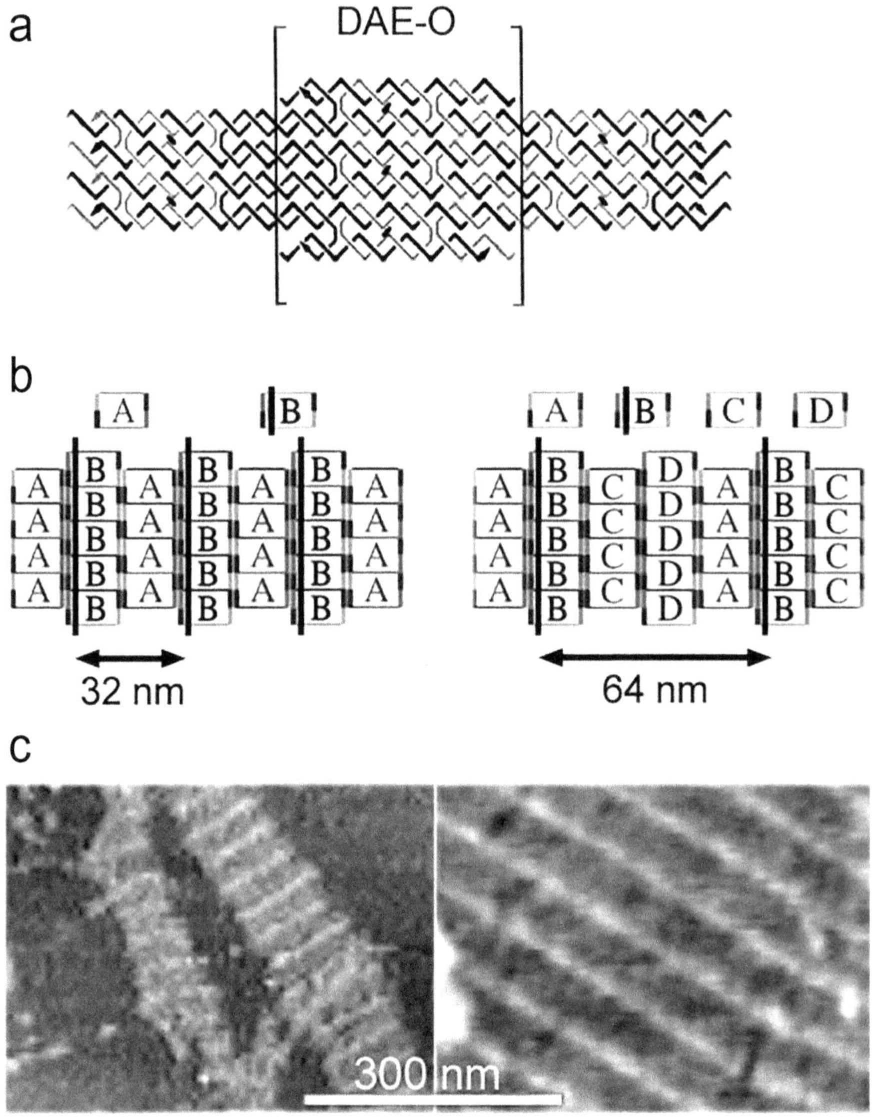

Figure 2.10 DX striped lattices. (a) DAE lattice topology with an odd (O) number of half-turns between crossings on adjacent units; one unit is bracketed. (b) Two structural types composed of two and four units; one unit carrying a hairpin is marked by a thick line in each type. (c) Their characteristic stripe patterns as seen by AFM. Reproduced from ref. 124 with permission from Springer Nature, Copyright 1998.

Another type of experimental design in which the introduction of periodic reporter strands can be effectively enhancing is used in ligation-closure assays. These biochemical methods are being progressively applied to the characterization of DNA nanodesigns (*e.g.* Borromean links).[81] In the case of DX elements, such assays can be used to assess the rigidity of the resultant structures.

In a typical experiment, the ligation products of DNA strands are examined to reveal if cyclization led to one product or many. The presence of many cyclic products would indicate that thc angles at the junction points of the tested molecule are not fixed.[81] Each of the oligomerized constructs contains an accessible reporter strand whose fate is identical with that of the whole complex. Therefore, topological consequences for the ligated DAE molecules containing five strands and the DAO motifs with only four would mean that only DAE molecules could generate a reporter strand.

However, the fifth, central, strand in DAE may not be able to close off. Therefore, an ideal solution to generating a reporter strand would be to extend it by introducing a bulged junction, that is, converting DAE into DAE+J. Both motifs contain five strands, two of which are continuous and three are crossover, including one central cyclic strand which in DAE+J forms a three-arm bulged junction. Ligating both DAE and DAE+J molecules generates reporter strands, resulting in negligible or no cyclization.[44] In marked contrast, a DAO motif lacking a reporter strand upon ligation gave a series of catenated molecules.

The main advantage of the double-crossover motif over single branched junction motifs, which cannot support periodic nanoscale configurations, is its structural rigidity.[9,85] This is a sought after criterion to ensure structural integrity in the assembly of periodic matter, which can be represented by crystal patterns.[8] The property can be used to specify sticky-ended associations for short-range constructs in a highly controlled manner. Since sticky ends are different, they are deemed asymmetric, which, as experienced in most nanodesigns, suggests that the best way of manipulating finite nanoscale structures is to minimize their symmetry. This should be taken as an adverse factor when attempting to construct longer range crystal-like objects for which symmetry is a major dominant feature. Indeed, the key distinction of crystalline materials is the translational symmetry of their units or motifs, suggesting that the motifs must be constrained to maintain the same spatial relationship with each other. Cyclization, knotting and, in some cases, catenation are the consequences of structural flexibility, which can be seen as assembly poisons for periodic structures.[80,81,125] It can therefore be concluded that the no apparent cyclization observed for the ligation closure of DAE/DAE+J motifs confirms the solution.

Further evidence can come from different edge-sharing DAE-based designs. In one, for example, it has been shown that the DAE motif furnished into one side (edge) of a DNA triangle maintains its structure, which allows it to oligomerize into micron-long 1D arrays with nanoscale periodicities,[126] Figure 2.11a.

The main objective of this design was to establish the compatibility of a rigid geometrical form, such as a triangle, with a rigid DX. The relative stiffness of the DX motif within the triangle was used as a parameter to assess ligation-closure cxperiments.[44,81] These experiments revealed that triangles with incorporated DX motifs along one edge could generate non-cyclic structures. This notion was further strengthened by AFM imaging, providing strong support to the drawn relationship between the model of ligation products and the observed zigzag assemblies, Figure 2.11a.

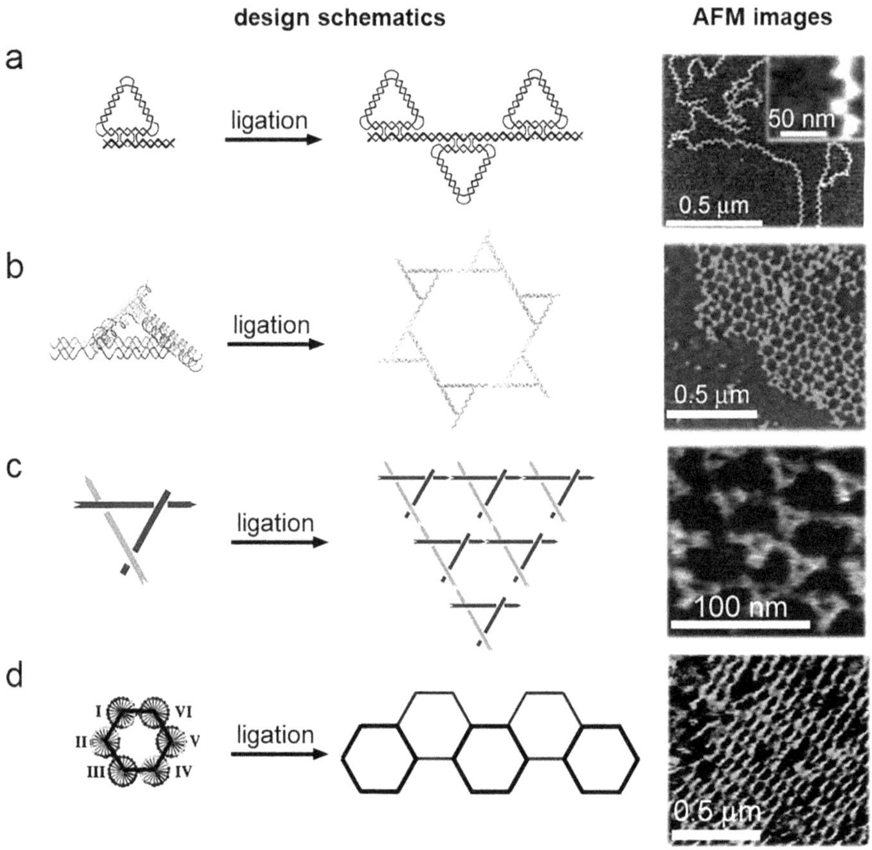

Figure 2.11 DX translational symmetry elements. (a) Ligation of a double-stranded triangle with one DAE edge into a zigzag 1D array. (b) DX triangle assembled into a pseudohexagonal trigonal lattice. (c) 2D triangle array built from an equilateral tensegrity triangle. (d) A ligation of a six-helix bundle (a cross-sectional view with numerically assigned helices) into a 2D hexagonal lattice. The corrugated arrangement is shown with the black layer being closer to the reader than the grey layer. Adapted from ref. 126, 127, 129 and 130 with permission from American Chemical Society, Copyright 1998, 2004, 2004, 2005.

It appears also that either DAE or DAE+J triangles can be used as building blocks for 2D hexagonal lattices.[44] To realize these, every triangle edge can be made to contain one DAE or one DAE+J molecule, giving different types of arrangements. If in the exclusively DAE-built triangle one edge is made of DAE+J, an extra helical domain will be produced within each hexagon. Instead, in the purely DAE+J lattice the extra junction is involved in the formation of triangles and it braces branched junctions to prevent them from bending, thus keeping them straight at their axes.[6]

However, the experimental feasibility of these arrays proved to be difficult to conform to the DX motifs, the rigidity of which could not match that inherent to triangles. The need to solve this problem stimulated the development of a new and more efficient motif – a DX triangle,[127] Figure 2.11b.

The element, the combined derivative of two previously engineered motifs – the double-crossover[45] and bulged triangle motifs[81,128] – has two key characteristics. The first characteristic is the fact that the DX molecules are twice as stiff as conventional duplexes,[67] and therefore the thicker edges of the formed triangle are expected to be considerably more rigid than those of the simple bulged triangular junction. The second concerns the innate property of the DX motifs for double intermolecular interactions as opposed to single helical interactions, which makes the designed triangle less prone to errors in twist.

These features, the synergetic combination of which all previous motifs lacked, were shown to be crucial in constructing trigonal arrays.[127] Two triangles were designed to co-assemble into a continuous DX structure with a pseudohexagonal lattice arrangement containing 13 double-helical turns (~46 nm), Figure 2.11b. The edges of each of the triangles were formed by 65 nucleotide pairs in each of their DX helices (four turns per edge), terminating with six-nucleotide sticky ends. Both pair- and sequence-wise the trigonal frameworks are identical and differ only in the sticky ends. A hexagon flanked by six triangles, demonstrates the arrangement, Figure 2.11b.

Because the edge of the hexagon lacks one triangle and has nine turns (~30 nm), the centre-to-centre distance can be estimated as ~34 nm. Consistent with the design, a characteristic honeycomb structure was identified by AFM, Figure 2.11b. As predicted, each unit of the structure has a hexagonal geometry formed by six triangles, with the centre-to-centre hexagon having an edge of ~38 nm. Importantly, to provide experimental evidence for the engineered structural characteristics of the DX triangle, namely doubly thicker sticky ends, the same set of experiments was repeated with the triangles having the sticky ends removed from one of the helices. As expected, no lattices were generated,[127] confirming that the use of double-helical arms is truly critical in promoting robust 2D grid-like assemblies.

Other systems and concepts were used to test this notion, which can be exemplified by the construction of stable tensegrity nanostructures, DNA equilateral triangles[129] and six-helix bundle motifs.[130] The first design relies on the integrity of connected assembly units rather than on their rigidity, Figure 2.11c.

The motif, built up of three four-arm junctions fused in a triangle, with each vertex formed by a four-arm junction and each edge by a duplex, assembles into 1D and 2D lattices, which consist of individual uniformly sized triangular particles. It appears to be possible to use the motif to guide DNA assembly in 3D, with all three components extending in three different directions, making it non-planar.

Three-dimensional arrays can be generated from motifs that are planar and rigid (such as DX), with their interhelical angles being flexible enough to permit small and fixable rotations. Therefore, 3D motifs that are inherently constrained rotations are more capable of setting up 3D structures, which was demonstrated using a DX version of the motifs – a 3D-DX triangle.[131]

The motif exhibits tensegrity geometry: the arrangement of a 3D triangular construction in which each side of the triangle is placed below one and above another.[131] This allows the design to incorporate crossover separations with even and mixed even and odd numbers of half-turns between all junctions. The repeat units contain seven and eight double-helix turns corresponding to roughly 25 and 28 nm, respectively. The motif self-assembled into extended and periodic 3D arrays, with repeat distances being in perfect agreement with the estimates.

The second design, 2D arrays assembled from hexagonal six-helix bundle DNA motifs,[130] can serve as an angle-related approach that carries major similarities to both the DX motif[45] and the tensegrity triangle.[129,130]

This motif, dubbed 6HB, was constructed from a series of DX motifs positioned parallel to each other at a specific angle such that the resulting structure would present a discrete bundle of six aligned helices interconnected in a fashion similar to that of DX motifs. Because the helicity of DNA is 10.5-fold, it was possible to assign crossover separations that would rotate the helices by 120°, Figure 2.11d.[132–134] The bundle can therefore be assembled in 1D, 2D or 3D: if the sticky ends of each helix in the bundle are made complementary to each other, 6HB molecules will organize into a 1D array, whereas 2D and 3D arrays are realized by bonding the helices in adjacent pairs.[135]

In short, pairing sticky ends on the front of helices I and VI in Figure 2.11d with the back ends of helices III and IV will give a 3D array (and the same for front–back pairs of II–III with IV–V and of IV–V with II–I), whereas a 2D array of alternating grooves and ridges is constructed when only two of the three pairs contain sticky ends.[131]

The outlined concepts can be expressed by other motifs based on similar geometric principles. However, "doubled" sticky-ended cohesion remains the main determinant. Employed in the context of sticky ends, it appears to have provided a compensatory mechanism for variations in connections between assembling components. Nanoscale 2D and 3D objects, which assemble *via* the point-to-point sticky ends, will still require geometric constraints, with different contact modes employed to carry this out.[47] For 3D structures, *e.g.* polyhedra, these can be corner-sharing modes, *i.e.* those that occur between vertices, or edge- or face-sharing, *i.e.* between two edges or all edges of two faces, respectively. Sticky-ended cohesion supports corner-sharing modes and, whereas face-sharing appears impractical, edge-sharing is prompted by the very architecture of DX motifs, Figure 2.12.

An extension of the sticky-ended assembly to edge-sharing was shown by designing two new motifs. These were constructed as DAE-fused modules shaped in an edge-sharing fashion as two and three DNA triangles, E2T and E3T, Figure 2.12a. The first, E2T, has two triangles linked by one DAE

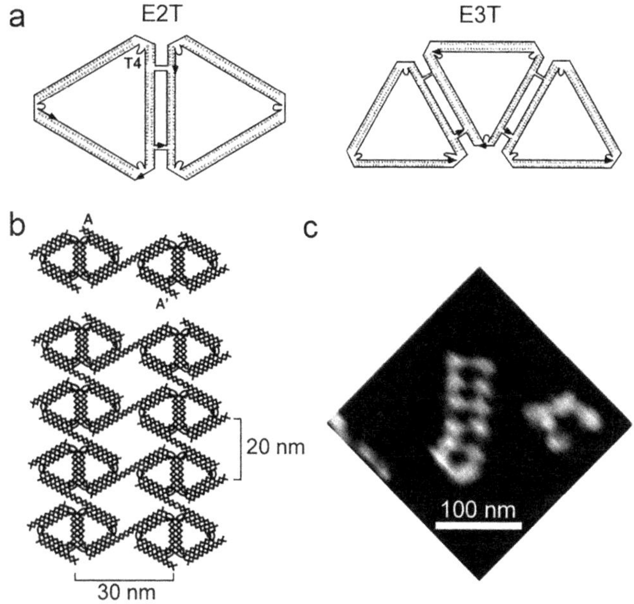

Figure 2.12 Edge-sharing DX motifs. (a) E2T and E3T motifs composed of two and three topologically closed triangles, respectively. Cyclic central and inner triangular strands hold the triangles together. The angles of each triangle are fixed by T_4 bulges in the inner strands. (b) Schematic of the double-row assembly of edge-sharing triangles. A double-row array of a conjugated version of E2T in which two units are covalently linked at one set of sites and blunt-ended at the other leaving only A/A′ sticky ends (top). The distance between the two fused rhombi is ~30 nm and the repeat distance in the array is ~20 nm. (c) An AFM image of the assembly. Reproduced from ref. 47 with permission from Elsevier, Copyright 2001.

construct. The molecule has four strands: an outside strand, two cyclic inner strands and another smaller cyclic strand involved in the two crossover structures. Three dT_4 bulges were introduced into each cycle to bend the corners, rendering the inner strands triangular, Figure 2.12a.

An extended version of the E2T design, the E3T motif, uses the same set of parameters, but is based on three triangles fused together by two DAE molecules, which gives a composition of six strands – an outside strand, three triangular strands bent by the same type of bulges and two circular strands. The outside edges of each triangle are three-turn-long double-helical domains. Both of these motifs are catenated structures. E2T adopts a rhombus-like shape and E3T is a half-hexagonal-like trapezoidal form.[47]

The motifs can assemble into different periodic arrangements. A DX molecule incorporated on each edge of E2T can give a construction with one domain of the molecule used as an extension for each outside edge and the second domain left to interact with a similar domain of another copy of the motif. Sticky ends direct interactions between the units of the motif. The four edges of the rhombus shape permit the assembly of 1D linear arrays using two sets of sticky ends. Moreover, two distinct types of arrays can be defined by different spacings between individual units of two different types. The units give two repeat necklace-like patterns with centre-to-centre distances of 20 and 30 nm, respectively. A conjugated version of the E2T motif can be designed to assemble by the sticky ends at one set of sites, *i.e.* A/A′, Figure 2.12b. The arrangement makes the distance between the two fused rhombi ~30 nm, and the two newly produced units give a double-row array with a repeat distance of ~20 nm, Figure 2.12b. Consistent with this, AFM imaging confirmed the formation of double-row 1D arrays with the distances between the units and rows matching those predicted, Figure 2.12c.

The concept can work for 2D arrays if its planarity is amenable in 2D.[136] Similarly, for 3D structures the edge-sharing concept may be extended to face-sharing.[36] As an exemplar, one can emulate the octahedral geometry of Buckminster Fuller's sphere to build discrete 3D structures. The sphere is a 12-connected octet truss structure presented by a series of fused tetrahedra. It is a stick version of cubic-close packing, which requires 12 arm junctions to build. However, in principle, the structure can be built as an array of face-sharing motifs. Other strategies offer alternative solutions to extending edge-sharing, which have already introduced a number of useful motifs such as paranemic[137] and triple[46] cohesion and square-aspect ratio tiling.[138]

2.3.3 Secondary Motifs Without Sticky-ended Cohesion

Paranemic cohesion is represented by a class of topologically closed complexes. These are coaxial four-stranded motifs termed paranemic crossovers, abbreviated PX.[137]

PX motifs are stabilized by every nucleotide paired by Watson–Crick bonding, which, in contrast to other four-stranded DNAs, eliminates the need for any other non-hydrogen bonding interactions, Figure 2.13a.

One PX molecule contains a central dyad axis, which relates two flanking parallel duplexes forming crossovers with crossover points flanking the axis at major and minor grooves. Two other helix axes flanking the crossover axis have a repeat which is similar to the normal twist in duplexes such that every strand in PX crosses over every helical repeat, but the structural period of each strand, and that of the PX structure, comprises two helical repeats. Thus, PX motifs are similar to and principally different from DX motifs in that both types are composed of two parallel helices, but crossovers in PX are formed at every possible point with no backbone juxtapositions, Figure 2.13a.

The pairing of two-component duplexes can be indefinite and occurs without physical linking between them, which makes the structure paranemic. The interdigitation of the duplexes can also be referred to as plectonemic because it is analogous to that of the strands in a double helix and is observed in supercoiled structures. Collectively, this brings up a very efficient mode of assembly that is alternative to sticky ends.

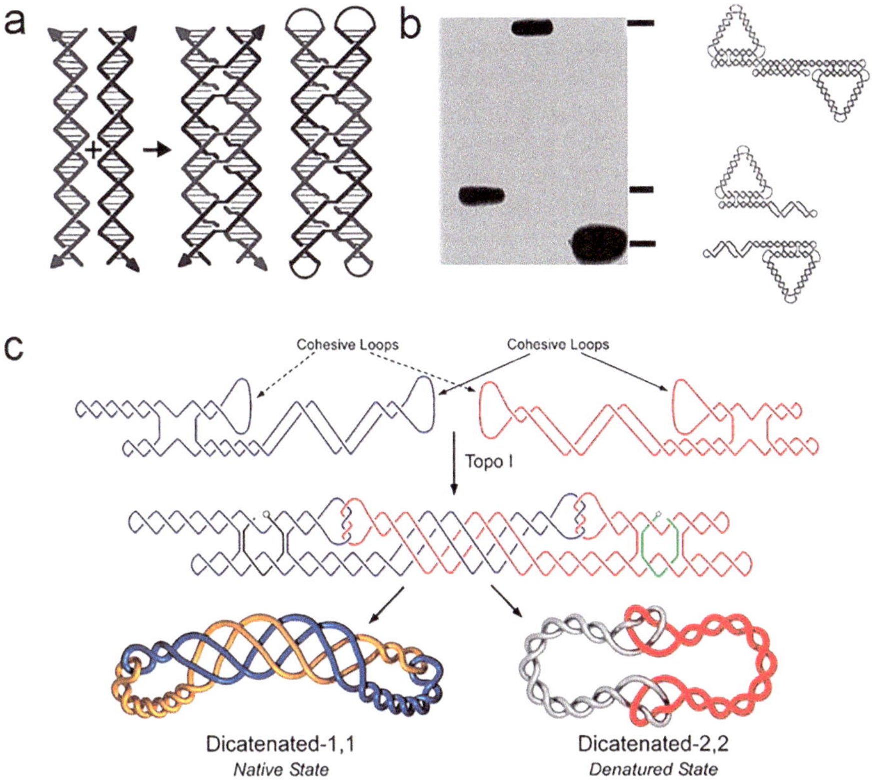

Figure 2.13 Paranemic crossover (PX) motif. (a) Formation of PX from two double helices exchanging strands at every point, resulting in the interdigitation of the helices. The helices in hairpin loops are capped to emphasize their paranemic relationship. (b) Paranemic cohesion based on two triangles with one DX edge each. Each tailed non-triangular DX domain produces PX with its complementary partner on the other triangle (right) corresponding to a specific band in non-denaturing polyacrylamide gel (left). (c) Schematic drawing showing PX cohesion of two circles with two sets of cohesive loops, annealing, ligation and knotting into PX-cohering tiles. Adapted from ref. 48 and 114 with permission from American Chemical Society, Copyright 2002, 2015.

A number of geometric designs[48,114,139] and sequence-dependent dynamic nanodevices[140–142] have been designed using this approach. An archetypal design is the 1:1 binding of large triangle motifs.[48]

The main consideration for this design is that sticky ends, which are normally introduced by restriction enzymes, are made very short to support complex large and stable structures. Assemblies based on long sticky ends are hampered by twist errors. Therefore, PX constructs can be designed as topologically similar to edge-shared[47] and DX triangles,[126,127] Figure 2.13a.

Each of two triangles is constructed with one edge presented by a DX molecule. As with edge-shared motifs, the two triangles are topologically closed molecules, but with the small central strand left unsealed. In contrast to those of the other two types of triangles, the triangle-linking extensions are complementary PX molecules, which gives the type of cohesion completed with the formation of a one-to-one structure. Evidence comes from gel electrophoresis experiments using equimolar mixtures of the two triangles with each containing a tailed non-triangular DX paired with the opposite segment of the other triangle by PX binding, Figure 2.13b.[48]

PX-cohesive molecules designed with complementary loops, which can be ligated by topoisomerase 1, proved to be instrumental in designing DNA topologies.[114] By varying the lengths of the complementary tracts between the loops and external conditions, it is possible to generate distinctive catenated and knotted structures, Figure 2.13c.

Another mode of assembly, which is as efficient an alternative to the sticky-ended cohesion as the PX motif, is triple cohesion.[46] This mode was introduced to accentuate the problem of coplanarity in building 2D lattices. A triple crossover, or TX, is a motif consisting of three adjacent helices, parallel and antiparallel, placed in the same plane, Figure 2.14a.

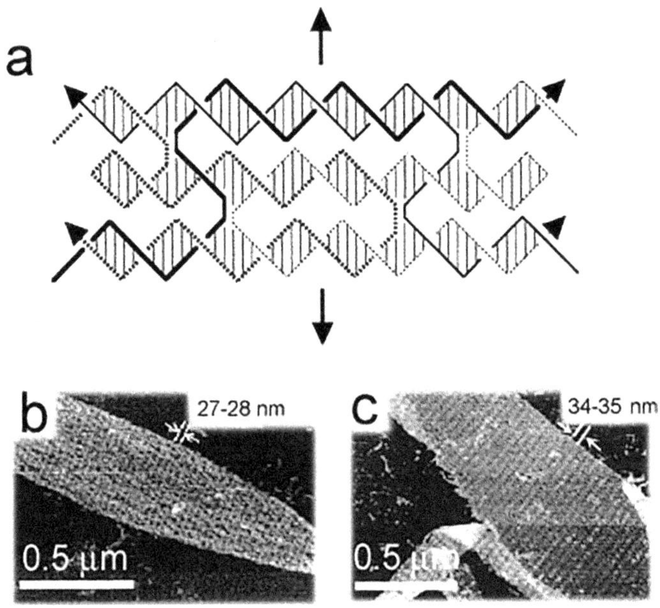

Figure 2.14 Triple-crossover (TX) motif. (a) TX assembly of four strands into three helices with coplanar axes. The middle helices are capped with hairpin loops. The two vertical arrows indicate the dyad axis, which interrelates two pairs of strands, shown as continuous and dotted. (b) 2D arrays assembled from two TX units. Visible cross-hatched stripes correspond to bulged reporters (TX+2J) uniformly separated at ~28 nm. (c) 2D arrays assembled from three TX units. Visible stripes caused by the capped outer helices of one unit with uniform spacings of ~35 nm. Adapted from ref. 46 with permission from American Chemical Society, Copyright 2000.

The helices are held by a minimum of two crossover sites between each of the domains rather than three crossover sites between two domains. The motif has all of its axes coplanar,[46] for which domains within each helical pair were made antiparallel. This is evident from Figure 2.14a, which shows that minor grooves abut major grooves in both interhelical regions. The molecule has four strands shared by three helices and four crossover points. Each vertical pair is separated by an odd number of half-turns: three between the left two helices and five between the right two helices. The central helix is capped with hairpin loops, Figure 2.14a. In this arrangement, TX appears as a double framework of two DAO components with a common central duplex. The design of this topology considers three important points.

The first point is that that motifs with contiguous patterns of reporter strands should allow for a more accurate acquisition of outputs in DNA-based computation.[8,143] The second is that the encapsulation properties of 2D lattices are desirable to accommodate guest macromolecules, *e.g.* in the assembly of molecular memory devices.[143,144] In contrast to DX constructs, a TX-based array is more intrinsic in generating a one-helix gap at a high density with just one species in the array.[8] The third point is to ensure large displacements in nanomechanical devices, which can exceed 10 nm when TX components are used.[46,140,145]

Two varieties of TX motifs for the assembly of 2D arrays provided experimental evidence support for these points, Figure 2.14b and c. In the first type, one TX module is coupled with a TX+2J motif – TX with two bulged junctions – to give 2D arrays with characteristic topographic features caused by bulged reporters. This is seen as uniformly separated cross-hatched stripes, Figure 2.14b. In contrast to DAE designs, in this design each component in the array is located between two 2 nm gaps, which can be used as encapsulation or binding sites.

In the second array type, three TX components were used to perform distance marking. Two, A and B, are typical TX molecules, with their central loops made sticky ended as opposed to capped by hairpins in their conventional counterparts. In the third component, C, outer helices are converted to hairpins, while sticky ends complementary to those of the first two are incorporated in its central domain. Because it is physically impossible to fit the third between A and B in the same orientation, component C is rotated about 103° relative to the location of A and B motifs. In this orientation, capped helices of C serve as topographic labels protruding from the plane, generating a uniform and predicted spacing.

The extended ligation of hairpin reporters, achieved at least within the 20 units range, in the formed arrays permits computational applications of TX-based arrays.[46] However, the main advantage of PX and TX motifs is the notion of multiple cohesion elements, which challenges the seemingly exhaustive role of sticky-ended cohesion, but also offers a strategy for mixed topologies. For instance, it has been shown that PX edge-sharing designed in a triangular form with four strands per edge can be used synergistically with the TX motif.[139]

In this design, two sides of one edge of the triangle were modified with two different sticky ends to bind one TX molecule, used as a mediator of the triangle assembly into a zigzag 1D array. Conceptually similar designs can include skewed TX triangle and double PX cohesion elements,[131] which can also be cloned.[37]

Single-stranded viral DNA, *e.g.* in M13 bacteriophage, is perfectly compatible with DNA nanodesign. The feasibility test of using it involved the replication of PX DNA.[146] The topology of the crossover could be generated from two closed dumbbells or a cyclic DNA, which required a DNA polymerase to "read through" the molecule to untangle all the crossovers before catalysing the synthesis. A moderate amplification of the motif provided evidence for the replication of artificial nanostructures, prompting the discovery of complex DNA motifs in natural systems.[37,146]

The motifs described to this end constitute a powerful toolkit in nanodesign. However, these motifs are largely based on a dyad symmetry, which is not devoid of shortcomings. Branch migration, *i.e.* an isomerization event accompanied by the relocation of branch points, is a relatively serious drawback with regard to controlling intended assemblies. This event is more evident in a DX tile when the structure is viewed as a collection of two four-arm junctions linked by two adjacent arms.[44] Although branch migration has to occur as a coherent process of both crossover points, the possibility of it increases with increase in the complexity of motifs and target architectures.[45] More stable immobile low-branched junctions can provide alternative motifs. Although these are excessively flexible at branch points, they inspire the search for geometrically more constraint crossover elements.

As a consensus, junction motifs with a number of double-helical arms greater than eight were proposed.[79] The geometry of these motifs permits the total exclusion of dyad symmetry and hence branch migratory formations. Since there are only four base pairs in conventional DNA, each of them can be used only once to flank a junction, hence junction motifs can be limited to four-arm junctions. However, with two identical pairs placed adjacent to each other, their pairing in the middle would be impossible. Therefore, junctions containing up to eight arms can be formed,[79] thus focusing on creating an entropic barrier for branch migration at the expense of the four base pairs arranged symmetrically about one junction centre. If applied to a higher arm junction, *e.g.* containing eight branches, the sequence of base pairs around the junction for arms 1–8 is arbitrary and can be made T–A, T–A, G–C, G–C, T–A, T–A, G–C, G–C. Repeating the same dinucleotide sequence every four arms effectively generates a 12-arm junction with the sequence G–C, T–A, C–G, A–T, G–C, T–A, C–G, A–T, G–C, T–A, C–G, A–T. Eight- and twelve-branched junctions constructed using such sequence symmetry revealed no branch migration.[79]

2.4 DNA Tiles

A promising solution to finding the best combination of the strong points of both element types has been formulated in the concept of a high square-aspect ratio.[138] This concept helps in the design 2D lattices with controllable encapsulation properties,[147] using a four-by-four tiling approach, which is the design of four four-arm junctions generated by nine strands, with one participating in all four junctions, Figure 2.15a.

When assembled into a tile the junctions become fixed to point in four different directions in the plane. To enable this, bulged loops were introduced into the shared strand to fold at each of the four corners inside the tile core. This also decreases stacking interactions between adjacent junctions, the inhibitory effect of which on the assembly was observed in the designs discussed earlier. The main constraint to each individual branch junction is provided by three others within the tile and a further one in an adjacent tile. The resulting structure is a rigid building block suitable for use in larger structures.

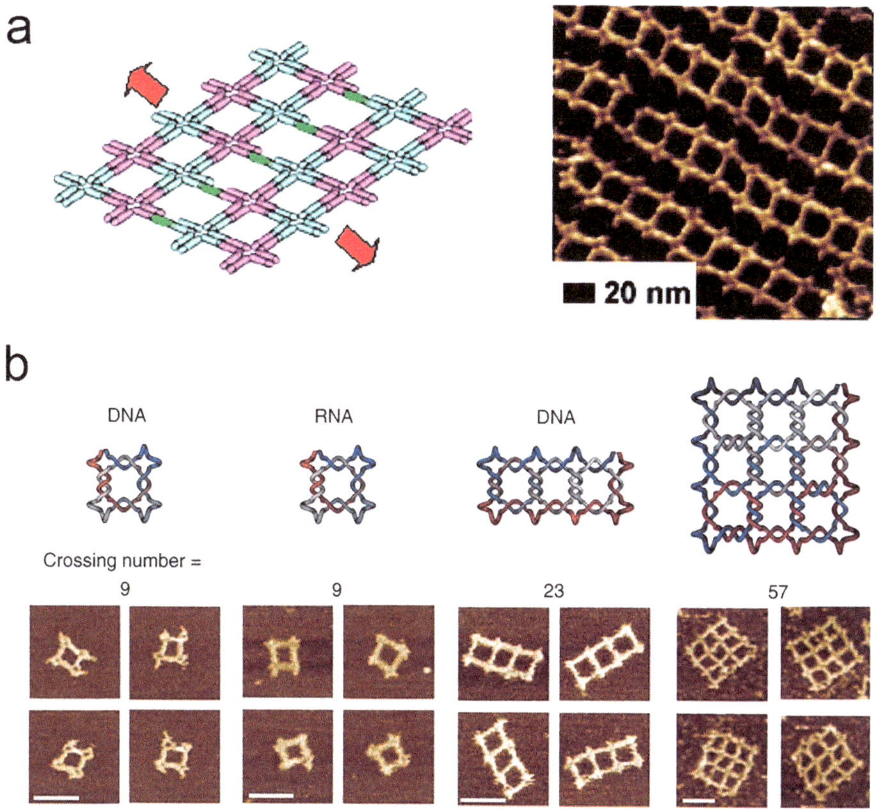

Figure 2.15 Design of high aspect ratio lattices. (a) DNA nanotrack, cartoon (left) and an AFM image (right). Arrows indicate the directions of growth at the expense of additional dsDNA bridges. Reproduced from ref. 149 with permission from American Chemical Society, Copyright 2008. (b) Design of 2D DNA and RNA knotted tiles with increasing crossing numbers (from left to right): Models (top) and corresponding AFM images (bottom). Scale bars are 50 nm. Reproduced from ref. 151, https://doi.org/10.1038/s41467-018-07039-7, under the terms of the CC BY 4.0 license https://creativecommons.org/licenses/by/4.0/.

Different assembly strategies were applied to control the periodicity of square cavities in the set lattice formation. In one design, the distances between neighbouring tile centres were made even numbers of helical half-turns, which allowed for the same face of each tile to point towards the same lattice face. The resultant structures were micron-long ribbon-like assemblies with uniform widths of about 60 nm.[138] The structures had two layers, in contrast to a flat lattice the edges of which are higher than the middle. The latter may indicate a finite curvature of tube-like lattices, which were predicted to result from each component tile being oriented in the same direction in the lattice plane, whereas incidental locally seeded curvature is extended over the entire lattice.

This proved to be the case in a design dubbed "corrugated". This design aimed to eliminate the surface curvature inherent for each tile within the assembly, which was done by inter-associating adjacent tiles with the same face of each tile being oriented alternately up and down in adjacent tiles.[138] Another objective was to produce larger pieces of 2D grids with square aspect ratio, which was ensured by making the distances between the centres of adjacent tiles four and a half helical turns. Combined with two double-helical diameters, this gives ~19 nm, consistent with AFM measurements. Large lattice pieces of up to hundreds of nanometres on each edge were observed with square-shaped cavities.

The most remarkable property of the assembled tiles is the flanking of the four-by-four arrangement cavities by segments of four separate tiles. This is in contrast to similar designs such as different types of parallelograms containing cavities defined by a single tile.[131,147,148] Such a multi-tile format prompts combinatorial approaches allowing binding cavities within the tiles and leading to specific patterns, which can be bridged or conjugated.[149,150]

This can be illustrated in two designs. The first design involves the stepwise assembly of DNA tile lattices *via* dsDNA bridges. As an assembly motif dsDNA is not that common, but it is appealing as a linker or a bridge connecting more complex supramolecular DNA structures such as cross-tile building blocks.[149] The use of dsDNA bridges helps control the length and directionality of tiles in a stepwise manner. It is also a means of reprogramming sticky ends displayed on the tiles, allowing for the modular assembly of fixed-size nanoscale arrays in addition to extended 2D crystals of nanoscale tracks, in which dsDNA serves as bridges, Figure 2.15a.

The other design is a protein-mediated synthesis of DNA crystals. The rationale of such designs is the use of proteins, which can manipulate DNA, as a means of controlling DNA assembly in addition to the hybridization and cohesion of DNA strands. RuvA, one of such proteins, was chosen to demonstrate the approach. This protein is a part of resolvasome, a protein complex resolving Holliday junctions during homologous recombination. The role of RuvA is concerned with holding the junctions in a square-planar configuration, promoting and resolving branch migration.

Holliday junctions can be used to build an array, with RuvA proteins binding to junctions and thereby directing the assembly of the array. This was demonstrated by comparing an array made of DNA only and another array made in the presence of RuvA. The addition of the protein changed the lattice symmetry and connectivity dramatically. The effect is notable, as immobile four-arm junctions are very flexible molecules with unfixed angles at branching points. They cannot assemble orthogonally unless the angle is fixed and can form only distorted or skewed structures such as Kagome lattices, as observed when DNA-only junctions were assembled. By contrast, RuvA insertions proved to help fix branches at roughly right-angles, promoting square-lattice arrangements.

Similar lattices and tiles are assembled from single-stranded DNA and RNA using the principles of DNA knotting described earlier. Such designs do not require bridging or protein-mediated control over the assembly of the resultant tiles, but instead rely on hierarchical folding in the prescribed order of carefully selected crossings. A generic design strategy has been offered to allow the assembly of 2D- and 3D-exhibiting complex topological features, with crossing numbers ranging from 9 to 57.[151] Moreover, the designed knotted tiles can be replicated enzymatically *in vitro* and *in vivo*, thus providing a generic strategy for the design and nanofabrication of complex nanostructures, Figure 2.15b.

To this end, it is clear that the time and place for expanding the portfolio of DNA motifs is favourable to move on to the challenge of constructing more complex and functional DNA materials. With all the progress made so far, an obvious question is what gaps and opportunities exist for DNA nanodesigns. This question is addressed in the context of programmable DNA hierarchies in the following section.

2.5 Programmed DNA Assembly

The described motifs hold demonstrable potential for the assembly of higher order nanoscale structures. Effective cohesion patterns, robust hybridization and complementarity can be designed to orchestrate binding interactions into virtually any assembly form. Incomplete interactions, however, may render complex nanostructures less probable or aberrant.

To minimize the impact of variabilities on the structures, versatile assembly modes are needed, which permits further design considerations, primarily concerned with the selection of specific geometries and symmetries. However, one determinant in this endeavour remains constant, namely the double helix – a molecular code of any nanodesign based on DNA. The code has been attested *via* discretionary algorithmic assemblies to programme DNA into a rich variety of scaffolds, shapes, sizes and morphologies, representative examples of which are discussed below.

A typical protocol in constructing relatively simple hierarchies such as knotted, catenated or tiled structures involves multiple reaction and purification steps. This approach is prone to assembly errors and structural variations, ultimately affecting the uniformity and monodispersity of the desired hierarchical assemblies. Limiting the number of steps is a logical and efficient way to mitigate this problem.

The first example of this strategy was to use a long single DNA strand as an assembly template, which is in contrast to the strategy of annealing several short strands into one nanostructure, as discussed so far.[152] This choice is also required to ensure geometric rigidity in the assembly, which effectively builds upon the use of branched junctions to arrange double-helix branches into continuously triangulated frames as a means of resisting longitudinal deformation and compression. Tetrahedral forms, which represent triangulated cages, give a mechanically resistant architecture,[153] whereas non-triangulated forms such as DNA cubes[83] and truncated octahedra[41] are easily deformed structures.

Based on these considerations, a single-stranded octahedron, the assembly of which requires rigid components, was designed using crossover motifs rather than immobile junctions. Twelve struts were used to build the octahedron. The struts constitute the edges of the octahedron connected by six flexible joints as its vertices, Figure 2.16.

The structure folds in two stages. In the first stage, a single strand of 1669 nucleotides, used as a heavy chain, binds together five light chains of 40 nucleotides each, giving rise to a branched-tree structure, Figure 2.16a. During this event, five DX struts are formed to give an intermediate structure with terminal half-strut branches, which are 14 unique loops with each being complementary to only one other loop.

In the second stage, conjugate terminal loops associate to form the remaining seven struts *via* intramolecular paranemic cohesion. The assembled octahedron thus contains five DX and seven PX components joined as six four-arm junctions, Figure 2.16b. The structure was confirmed by cryo-electron microscopy and single-particle reconstruction to reveal octahedral objects of the target shape and predicted size, Figure 2.16c.

Importantly, the designed folding was not accompanied by the formation of knotted or catenated regions, suggesting that no covalent bonding was required for the generation of the octahedron. This, in contrast to the previously reported truncated octahedral and cubic structures, effectively demonstrates a single-step construction protocol based on a single folding event, which could permit the reversible strategy in the assembly of 3D DNA structures. The design also explicitly shows that a triangulated architecture provides discrete 3D objects with the necessary deformation resistance. However, proportionally high ratios of misfolded and distorted octahedra could still be observed in the assembly, suggesting that rigidity should not be the only factor required to support discrete 3D assemblies. In addition, it does not relate to other possible complications such as relative stability and steric purity of target structures. Hybridization interactions require a degree of stereospecificity towards one desired diastereomeric form, prompting the development of approaches that allow for a chirally specific assembly.[153,154] A stereospecific assembly of DNA tetrahedra was demonstrated to give a single diastereomer with 95% yield. The resultant assembly consists of rigid triangles built from four

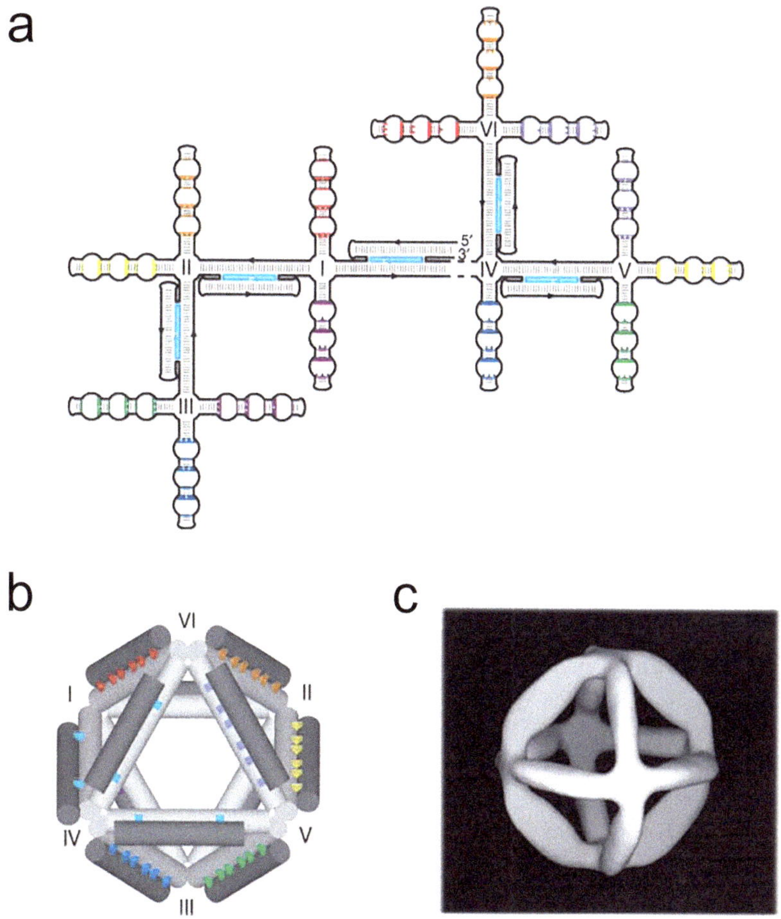

Figure 2.16 Design of a DNA octahedron. (a) Branched intermediate assembled from a single heavy chain (black) and five unique light chains seeding five DX struts (cyan). Complementary half-PX loops associating into seven other struts are coloured alike. Roman numerals indicate six four-arm junctions linking the core-layer double helices of each strut. (b) 3D folding of the intermediate into an octahedron. Twelve correspondingly coloured struts make up the edges connected by the six joints forming vertices. (c) 3D map generated from single-particle reconstruction of the octahedron. Reproduced from ref. 152 with permission from Springer Nature, Copyright 2004.

single strands, Figure 2.17a. When annealed, the strands assemble into one tetrahedral face, each forming double-helical edges of triangles covalently linked at their vertices. Four edges have single-stranded breaks, or nicks, and each vertex was closed through single unpaired "hinge" bases. When combined in an equimolar ratio under hybridization conditions, the strands assemble into tetrahedra in a sequence-specific manner with

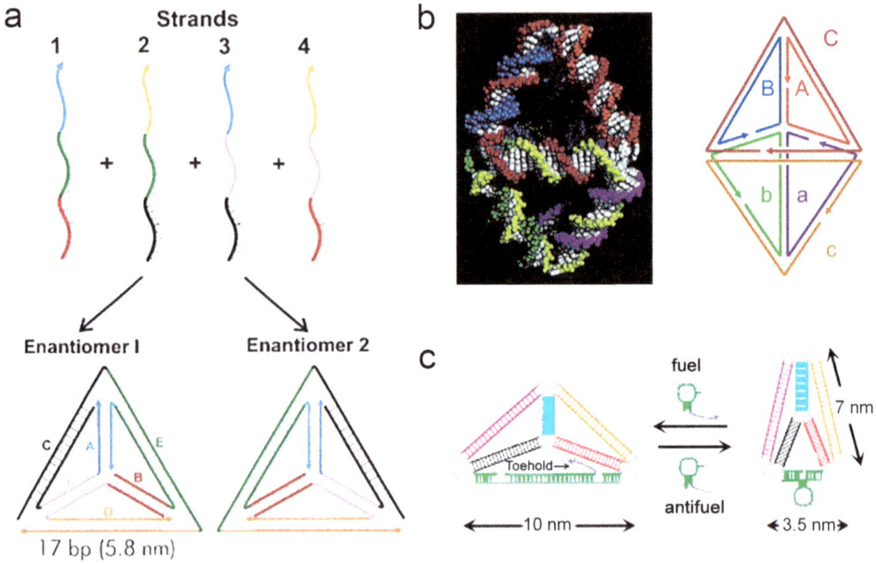

Figure 2.17 Design of DNA tetrahedra. (a) Self-assembly of four single strands containing complementary subsequences (coloured alike) forming the edges of a tetrahedron. The resultant tetrahedra can be one of the two possible enantiomers. Reproduced from ref. 154 with permission from the Royal Society of Chemistry. (b) Left, molecular model of a DNA trigonal bipyramid; right, positions of six constituent strands (a–c and A–C). Reproduced from ref. 155 with permission from American Chemical Society, Copyright 2007. (c) The "closed" tetrahedron (right) is opened (left) upon the addition of a "fuel" hairpin strand and is closed again by adding the "antifuel" hairpin displacing the "fuel" strand by hybridizing to the toehold region first. Reproduced from ref. 156 with permission from Springer Nature, Copyright 2008.

nearly quantitative yields. This can be followed by enzymatic ligation of the nick sites to close the structure into a catenane. Each edge of the tetrahedra has a different sequence, which renders each tetrahedron chiral, with the two possible enantiomers formed, Figure 2.17a.[154]

Such tetrahedra, assembled in a single step, can be used as building blocks for more complex 3D structures. This was demonstrated by bracing different tetrahedra at their gapped edges with programmable single-stranded DNA linkers. Strands linked with two identical or different subsequences could be designed to connect preassembled tetrahedra into homo- and heterodimers, respectively. The dimer formation can then be used to investigate the stereoselectivity of the assembly.[153] Namely, the position of a gapped edge of one tetrahedron determines the azimuthal orientation of the free end of a linker and by this whether it can hybridize with another tetrahedron. This allows the design of two gap positions, one for each of the two strands forming the edge, and hence two configurations.

In one configuration, the linking strand would project away from its centre, but in the other into the centre, which is expected to be inaccessible.[153] The results of linking experiments are consistent with producing the diastereomer having the major groove of each helix facing inwards at each vertex. This reflects a significant difference between the stabilities of the two possible diastereomers. The rigidity of tetrahedra assessed in response to axial compression was reversible for forces up to 100 pN while undergoing double-helix buckling under higher loads. With such a combination of properties – stereoselective synthesis and structural rigidity – the design conveys a facile route to defining relative coordinates of any part of a more complex 3D geometry with near-atomic precision.

More complex structures could also be accessed by this design strategy. For example, a larger but less symmetric polyhedral nanostructure was demonstrated in the assembly of a trigonal bipyramid.[155] In this design, the same single-step annealing procedure was employed to assemble six DNA strands, Figure 2.17b. Each of the strands, by running around one of the faces in a way analogous to the original tetrahedral design, hybridizes to an adjacent strand with the formation of nine edges of ~7 nm each (two turns of the double helix). Nicks were introduced into some of the strands, and single unpaired nucleotides were used as joining hinges. The bipyramid appears as a braced set of triangles formed by three four-arm and two three-arm junctions. Upon ligation of the preassembled structures, nine pairs of strands designed to share one edge and the six pairs of strands designed not to share an edge convert to linked and unlinked circles, respectively. The formation of catenated structures found in 15 different bipyramids confirmed the topological relationship between each pair of strands as designed. However, ligation of more than two strands can give rise to more complex catenanes and can be associated with the formation of by-products. This prompted another concept aimed at making DNA tetrahedra reconfigurable in response to specific molecular signals.[156] The ability to control the shape of a nanostructure may lead to the development of a nanoscale machinery operational on a single-molecule level.

To probe this, a tetrahedron with an additional structural module, reconfigurable edge, was designed. Similarly to the two previous designs, the object was assembled from four short strands each running round one side of the tetrahedron. Six pairs of complementary regions form six edges with each strand containing three of these domains. However, in this design the tetrahedron was made uneven in all of its edges: five of the edges were ~7 nm in length and the sixth ~3.5 nm, Figure 2.17c.[156]

Hence the key feature of the design is the insertion of a hairpin loop into the centre of the strand, which is opposite to the nick of its complementary strand. The loop is meant to act as a reconfigurable motif such that the tetrahedron would be able to adopt "closed" and "open" states as driven in response to the configurational change of the loop. When the motif folds as a hairpin, the edge it is inserted to is at its shortest and the tetrahedron is closed. The extension of the edge by a "fuel" strand, that is, the oligonucleotide that opens the hairpin by hybridizing to both halves of the stem and to the loop, gives a stretched or "open" form of the tetrahedron. The opened edge thus becomes longer and is a continuous 10 nm duplex with two nicks. The fuel strand contains a 3.5 nm extension, which within the formed tetrahedron remains unhybridized, giving rise to a toehold, Figure 2.17c. This can then be used to promote the displacement of the fuel strand by an "antifuel" strand. Because the latter is complementary to the entire fuel strand, including the extension, its hybridization propensity is higher and can be used to reverse the hairpin to its original closed state. Therefore, the sequential addition of the strands can change the length of the reconfigurable edge by a factor of three, repeatedly. Such specificity of the reaction permits the design of a four-state tetrahedron incorporating two hairpin loops in opposite edges. The loops in this structure are topologically similar but are assembled from different sequences. Engaging the two edges in an independent manner was shown to promote the switching of the tetrahedron between four different states with all possible combinations of open and closed edges.[157] It was possible, for instance, to convert the tetrahedron from an entirely closed to an open state by opening either hairpin or both sequentially.

The designs represent a DNA geometry dynamically controlled at the nanoscale. This makes it potentially useful for actuating single molecules or discrete assemblies on nanostructured templates or making 3D DNA shapes robotic, *i.e.* reversibly operational in 3D. The experimental validation of such applications relates to the problem of scalable DNA assembly and that of a single-step, self-reproducing automaton capable of translating algorithmic inputs into the construction of an arbitrary object.[158]

It can be argued that the notion of DNA automata was seeded by pioneering concepts of DNA computing and tilling.[12,70,159] The concept originated as a theoretical attempt to provide a link between DNA self-assembly and computation and sets out to demonstrate the compatibility of DNA self-assembly with Turing-type computers, finite-state machines that by moving back and forth on an infinite 1D memory tape can perform any computation, which is otherwise performed by other mechanical devices of any complexity. This makes it tempting to explore the theory and apply it to prescriptive DNA assembly.

It was originally hypothesized that a natural Turing computation model implemented by 2D algorithmic self-assembly would be the class of 1D or local cellular automata.[160] The binary exclusive disjunction or exclusive-or (XOR) function was chosen as an update rule for local automation, allowing each cell to be computed as the XOR of its two neighbours at each time step, and as it grows a fractal pattern, *i.e.* a Sierpinski triangle.

This implies that to perform a cellular automation using molecular self-assembly, a similar pattern with an intrinsic binary function should be employed. The formation of tiles, *e.g.* triangular patterns, by sticky-ended cohesion, as a binary function, is deemed obvious in this regard. The assigned program is not a sequence of instructions but a 2D aperiodic framework of tiles assembled from different pieces *via* sticky ends.

However, because molecular self-assembly is asynchronous and prone to errors, four main challenges must be solved for the experimental validation of the chosen cellular automata.

The first challenge, defined as the translation of abstract tiles into molecular tiles, which can form 2D crystals, is solved with the help of previous designs of highly periodic tiled crystals.[46,124,147] Two other challenges – finding specific binding domains from which molecular tiles can be assembled cooperatively to minimize errors and according to the logic of the chosen type of abstract tiles – is addressed by sequence-specific hybridization patterns, which were introduced to support addressable inter-tile interactions.[124,161] For the final challenge – input assembly of molecular tiles with the exclusion of spurious nucleation – the answer is the design of single-stranded templates capable of directed nucleation maintaining complex lattice-type assemblies, *e.g.* barcode patterning.[46,162]

The designed strategy was confirmed experimentally in the translation of an abstract pattern into DNA tiles derived from DX motifs and long single strands used as a tile-nucleating input. The patterns produced contained 100–200 correct tiles assembled with low error rates, which indicates that all four features for Turing-type universal computation can be achieved experimentally with high yields and accuracy. Controlling nucleation of crystal growth *via* selective and cooperative tile associations directed by sticky ends into the formation of extended crystals shows that DNA can be expressed in a desired algorithm and for any construction target.

2.5.1 DNA Origami

An approach to fully realize programmable DNA designs was devised as a "scaffolded DNA origami". This is an elegant one-pot method using ssDNA strands to direct the folding of one long single DNA strand into any 2D or 3D shape. In assembling DNA origami, a computationally assigned layout of a scaffold strand, a long single-stranded DNA, was used to fill a specific pattern row by row. To fix and rigidify the strand within the pattern, staples, short ssDNA molecules, were introduced to cross over from one row to another, hybridizing with the corresponding complementary regions of the scaffold strand.[13] In practical terms, the design required a sequence of five steps, as follows.

In the first step, a geometric model of a desired shape with specified sizes is built. The shape is then filled from top to bottom with an even number of parallel helical duplexes that are cut to fit the shape in sequential pairs. These are constrained by being an integral number of turns in length, thus requiring the fixing of the helices together by a periodic array of crossovers to mark locations at which strands running along one helix switch to an adjacent helix and continue from those points.[13]

In the second step, the scaffold strand composed of about 900 nucleotides is introduced to fold back and forth in a cellular pattern such that it provides one of the two strands in every helix. This introduces a raster progression from one helix to another, generating a set of crossovers, termed "scaffold crossovers", with the scaffold able to form a crossover only at those positions where the DNA twist places it at a tangent point between helices. This in turn implies that to transfer the progression from one helix to another and onto a third, the separation between successive scaffold crossovers must be an odd number of half-turns. However, where the direction is reversed vertically returning to a previously visited helix, the separation must be an even number of half-turns.

In the third and fourth steps, the designed model and path are represented as lists of DNA lengths and offsets in units of half-turns that together with the sequences of the scaffold and staple strands are input to a computer program. The final two steps, the fourth and fifth, are rectifying. The fourth is in that it introduces minimization parameters to balance the twist strain between crossovers imposed by the asymmetric nature of the helix, *e.g.* major and minor grooves, non-integral number of base pairs per half-turn. Namely, periodic crossovers are arranged with glide symmetry allowing minor grooves to face alternating directions in alternating columns of periodic crossovers, whereas to balance scaffold crossovers their twist is recalculated and their position is changed, typically by one base pair, while recomputing staple sequences. The fifth is in that it introduces an additional constraint parameter to strengthen the binding of the staples with the scaffold strand. This is stimulated because the staple strands at their meeting positions leave nicks in the backbone. By merging adjacent staples across nicks, fewer, but longer, staples with larger binding domains were generated.

Additionally, as the designed folding path compatible with a circular scaffold leaves a "seam" contour which the path does not cross, another pattern of breaks and merges may be created to cement the seam. In this case, a seam crossed or spanned by merged staples would be "bridged". Thus, when used in combination, the steps can give rise to a single algorithm allowing one-pot synthesis of an intended DNA shape. Figure 2.18a sums up all of the design steps.

To provide empirical evidence for the algorithm, a modified version of the 7 kb circular genomic DNA of an M13 phage was chosen as the scaffold strand. Not all segments of the obtained strand could be used and therefore remaining strands were added to complement unused sequences. In a typical experiment, staple and remainder strands are mixed (200–250 strands in a 100-fold excess) with the scaffold strand. The complex is then annealed and imaged by AFM.

Six different folding patterns were reported to demonstrate the method, Figure 2.18b. In the simplest and most flexible structure – a 26-helix square with no vertical reversals – a linear scaffold with a 2.5-turn crossover spacing was used together with 26-mer staples bound to each of two adjacent helices. Over 10% of observed structures were well-formed squares, with the remainder being deformed rectangular fragments.

To probe the formation of a bridged seam, a rectangle was produced using a 1.5-turn crossover spacing, 32-mer staples and a circular scaffold.

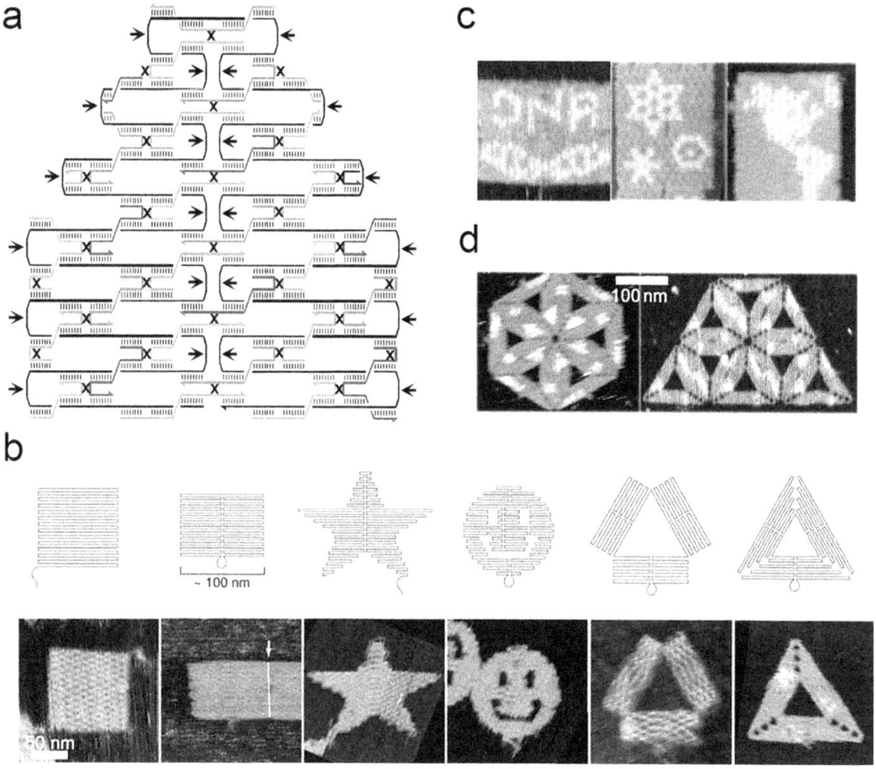

Figure 2.18 DNA origami. (a) A shape created by parallel double helices assembled from a scaffold strand (black) running through every helix and joined with staple strands by periodic (×) and scaffold (black arrows) crossovers. (b) Different folding paths and their corresponding AFM images: square, rectangle, star, disk with three holes, triangle with rectangular domains and sharp triangle with trapezoidal domains (from left to right). Hanging tails and loops denote unfolded sequences, the white line and arrow indicate blunt-ended stacking and white features (sharp triangles) indicate hairpins. (c) AFM images of assembled patterns: a pattern representing DNA with a pixelated turn of ~100 nm, a hexagonal pattern reflecting a nearly hexagonal pixel lattice and a map of the western hemisphere with a scale of $1:200 \times 10^{12}$. (d) Assembly of sharp triangles with extended staples into hexagons (left) and lattices (right). Reproduced from ref. 13 with permission from Springer Nature, Copyright 2006.

The central seam in the rectangle assembled with nearly quantitative yields was readily formed and could be easily visualized. The applicability of the algorithm to the construction of arbitrary shapes was exemplified by designing a five-pointed star with a 1.5-turn spacing, 32-mer staples and a linear scaffold. Similarly to the square, the design returned 10% of well-formed star-like structures. In both cases, the low yields of assembly could be attributed to strand breakages during syntheses involving enzymatic reactions and subsequent AFM processing. Indeed, when a circular scaffold was folded into stars, the yields improved to over 60%.

The third form demonstrates the assembly of scaffolds into different shapes without the need for topological disks. By contrast, a three-hole disk, a smiley face, could be constructed assuming a highly asymmetric folding path of five distinct seams. This design also revealed a tendency for deformations, while producing 70% of well-shaped structures.

In an attempt to combine raster-fill domains in a non-parallel arrangement, a triangular shape was designed comprising three separate 2.5-turn spacing rectangles, which were held together by only single covalent bonds along the scaffold. The target equiangular triangles appeared to be scarce and also tended to undergo stacking by rectangular domains, which could be attributed to the flexibility of single bonding at the vertices. An effective solution to this imperfection was found in a sixth form, the so-called sharp triangles, which were built from trapezoidal domains with 1.5-turn spacings. The domains were designed with slanted edges, which allowed the trapezoids to meet at the triangle vertices, generating more extended interfaces bridged by additional staples. The resultant well-formed triangles were more abundant compared with just 1% for the rectangular triangles. Yields of more than 80% for triangles bridged at the vertices and 50% for unbridged structures were observed.

The success with the outlined designs prompted a hypothesis that staple strands can mediate the decoration of shapes with arbitrary patterns of binary pixels, which can be done through labelling DNA with small molecules such as biotin or fluorophores.[13]

If the original set of staples is viewed as a binary "0" then a new set of labelled staples, *i.e.* one staple for each original, can give a binary "1". Thus, any pattern can be created by mixing subsets of these strands. A set of hairpins was designed to decorate the middle of staples at the position of merges. The resulting pixel patterns were found to depend on merge patterns and to be rectilinear, *i.e.* with adjacent columns of hairpins on alternate faces of the shape, or staggered, *i.e.* geometrically packed with all hairpins on the same face. When imaged by AFM, labelled staples as expected gave greater height contrast than unlabelled staples, 3 nm against 1.5 nm. This resulted in a pattern of light "1" and "0" pixels. Some patterns, each with 200 pixels, constructed using these conventions are shown in Figure 2.18c.

Similarly, extended staples were used to produce combinations of shapes. Such staples connect shapes along their edges, which is done by merging and cutting normal staples along the edges. Starting with the sharp triangles, Figure 2.18b, this approach was used to create finite and periodic structures, Figure 2.18d.

For the design perspective, this method is unique in that it ignores the need for careful sequence optimization and synthesis, which may be because DNA assembly is defined by a single parameter – long single-stranded DNA. It may not be possible to construct smaller scaffolds based on scaffold strands of considerably shorter hybridization segments as these would result in assembly variations. The method uses an arbitrary one-pot synthesis, which may be exclusive for building large 2D DNA scaffolds whose construction principles remain less amenable to lower-scale systems. DNA origami has proven to be an extremely attractive approach for assembling discrete extended arrays capable of mediating various macromolecular arrangements, including 3D shapes. When it comes to the designs of new forms and technology applications using the principles of DNA origami, the numbers are indeed staggering. The method of DNA origami has undergone further developments over the last decade, in which optimization challenges such as size limitations, issues with stability and scale of production have been identified. Empirical solutions are being proposed to the point of developing next-generation techniques enabling *in vivo* origami synthesis and multiscale manufacture.[163]

An obvious application of the method is the creation of a "nanobreadboard",[13] which can serve as a template for seeding different bio- and physicochemical components in a highly controlled manner. The idea of using arrays with nanoscale periodicities contributes to approaches to nucleate protein crystallization, template protein assemblies, build enzyme factories and obtain multilayered and extended matrices.[163–166] However, the field of 3D functional structures is where DNA origami is likely to impact most. A growing number of designs are being reported, although most of these are focused on a particular function or operation, without expanding beyond a few forms. DNA nanotubes represent one such form.

2.5.2 DNA Nanotubes: Intermediate Scaffolds for 2D and 3D Systems

Strictly, nanotubes are non-planar 2D arrangements with the characteristics of the third dimension. They may serve as an intermediate type of DNA scaffold for 3D systems that are expressed in and employ the dimensions in their architectures.[6,9]

Nanotube designs proposed and characterized to date can be loosely viewed as closed or wrapped tiles. In fact, the first nanotubes were accidentally encountered in tile-based assemblies[124,130] when it was suggested that the tiles are intrinsically inclined to form narrow micron-long ribbons by curling and closing on themselves, which appeared to be consistent with the fact that the tubes exhibit helical order.[167–169] The use of secondary DNA motifs in conjunction with 2D tiling provided an empirically promising strategy. The main challenge was to establish a set of rules linking the design of tiles with their tubular assembly in a predictable manner. The first nanotube tiling design relied on the apparent consistency of tile symmetry with the tube-type assembly and morphology.[138] However, the relationship between the tile curvature, geometry and spatial selection in tube formation was not entirely clear. As a consequence, a variety of tube sizes mixed with fragmentary lattices were observed for the design, likely indicating lack of control over the desired forms. A covalent constraint can be introduced to rectify this, serving as an orientation marker.[167] This strategy aimed to convert flat tiles into uniform nanotubes using thiols introduced to dsDNA loops projecting out of the tile plane. This is meant to render the systems switchable at the expense of reversible disulfide bond formation. As judged by a series of microscopy measurements, nanotube structures were indeed the only observed products resulting from annealing thiol-modified strands.

The design featured two tile building blocks. Both were based on the TX motif, namely TAO (tile A) and TAO+2J (tile B with two bulged loops), that is, as formed by three coplanar double-helical domains (T) with strands reversing direction at crossover points, hence antiparallel (A), having an odd number of helical half-turns between the points (O). The two bulged loops in tile B protrude one into and one out of the tile plane. All B tiles were modified by replacing one loop with two thiols coupled one on each of the termini. Sticky ends were placed at the corners of the tiles to allow two types of tiling. Nanotubes are formed as a result of alternating assembly of B and A tiles. In this mode, the dsDNA helix axes are parallel with the tube axis, and thiols are inside the tubes. The resulting structures were microns in length with uniform widths of about 25 nm.

The hypothesized burial of disulfides within the tubes as an indication of disulfide formation between adjacent B tiles – the force that causes the formed lattice to curve into a tube – was confirmed using thiol-reactive colloidal gold, which failed to detect any thiols on the outer surface. This provides sensible evidence that the structure is formed periodically by an even number of tiles per ring layer, namely eight tiles per layer, and that the rings are flat. A potential drawback of this design is the lack of generality in constructing DNA-only systems without the need for additional modifications.

With this in mind, a set of general design principles allowing the assembly of DNA nanotubes was devised by introducing a DNA-only nanotube motif.[170] A proposed motif is composed of two parts. One is a core represented by DX motifs and containing five pseudo-knotted strands that form two double-helical domains fixed by a pair of crossover points. Two different cores assigned by two different sequences were named RE and SE. The second part is the two pairs of ssDNA sticky ends flanking the core, thus allowing it to bind other cores, Figure 2.19a.

A single core yields a single tile able to assemble into a lattice independently or in combination with another tile type, which can be done by assigning specific sticky-end codes, Figure 2.19a. Lattices may be constructed with stripes running either diagonal or perpendicular to the long axis of the tiles. This can be accomplished, for example, by using reporter hairpin structures formed on the outside of nanotubes in a way similar to the previously described approaches. A collection of tiles is thus generated using a particular pattern, which is set by a particular choice of sticky ends. In order to drive the assembly of such patterns into tubes, a lattice should exhibit an appropriate symmetry, which was chosen as a rotational symmetry compatible with an intrinsically curved geometry for a patch of tiles.[170]

The symmetry confers necessary non-zero curvature and promotes tube formation, which is also mediated by the orientation and spacing of the crossover points. These identify factors underpinning the tube formation, with most of them met by an antiparallel DAE type of DX motifs having an

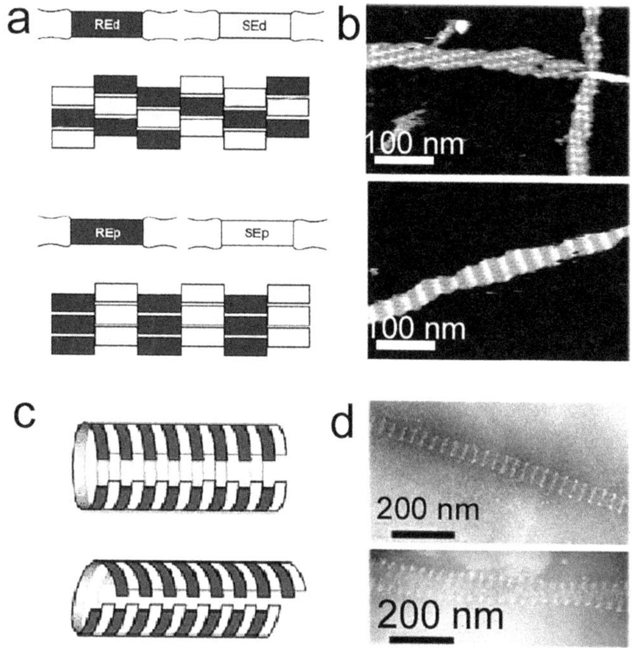

Figure 2.19 Design of programmable DNA nanotubes. (a) Schematic representation of DNA tiles with two different DAE-E cores, RE and SE. In combination the tiles can assemble into lattices with diagonal, REd+SEd, or perpendicular, REp+SEp, stripes generated by reporter hairpins (not shown). (b) AFM images of the diagonally (top) and perpendicularly (bottom) striped open nanotubes. Reproduced from ref. 170 with permission from American Chemical Society, Copyright 2004. (c) Folding of 2D arrays assembled from two types of tiles (indicated in grey and white) into tubes with alternating rings (top) and nested helices (bottom) of the tiles. (d) TEM images of correspondingly striped nanotubes. Electron-transparent stripes indicate protein-labelled tiles. Reproduced from ref. 169 with permission from American Chemical Society, Copyright 2004.

even number of half-turns between intermolecular crossovers, *i.e.* DAE-E. Other antiparallel DX types that have spacings with odd numbers of half-turns were therefore considered incompatible with tube formation. Basic features of the resulting model were confirmed by AFM of nanotubes derived from different tile sets, showing tubes with diagonal and perpendicular stripes, Figure 2.19b.

In most cases, the tubes were closed single-walled tubular structures able to open into sheets. The model assignment of the inside and outside of the tubes was tested assuming that large decorations on the inside of a given tube would lead to disruption of the tube formation, which was largely confirmed. For tiles with hairpins predicted to lie on the outside and inside of tubes, the formation of morphologically normal tubes was observed only for tiles with the outside hairpin insertions. In the other two, the tube formation was disrupted, leading to variable morphologies from short open to flipped tubes and their combinations. Collectively, the experiments provided a good correlation for the proposed close-packed model of a heptagonal tube.

Using a similar strategy, it proves to be possible to design a DX-based tiling system capitalizing on the chiral nature of the double helix leading to the formation of chiral ribbons with straight edges. The chirality of the nanotubes results from the pattern of connections between two types of tiles designed to tessellate by assembling as alternating rings or nested helices, Figure 2.19c. The ring and helical order of the tiles was confirmed by biotin–streptavidin labelling, which helped to reveal straight and spiral stripes across the assembled tubules, Figure 2.19d.

These examples are based on structural parameters that primarily stress on rigidity provided by crossover motifs and curvature-promoting symmetry in tiling. However, a high occurrence of morphologically polydisperse tubes in such designs stimulated the development of more robust theoretical frameworks.

In pursuing this, a classification model was proposed to design minimally strained nucleic acid nanotubes, simplifying design processes by adapting certain types of symmetry.[168] This approach implies that most nanotubes are rigid regular polygons, which may be a limitation to nanotube design, but allows parameters for the classification of nanotubes to be determined. Approximating this to the most stable and minimally strained structures modelled against constraints characteristic for DX motifs was suggested as a perfect fit of double-helical structures to a set of duplex parameters. With a number of further approximations allowed by such parameters, classes of low-stress DNA nanotubes, which defer by a number of helices and cavities of varied shapes, were predicted. Although the approach does not consider strained forms, it may be used to distinguish secondary and tertiary structures, particularly in the context of primary and secondary motifs.

As an experimental exemplar for the modes, a hollow tubular system was designed to partially cover and then sheath its content.[171] This is in contrast to a more common approach of delivering target entities into the preformed cavity. The design uses DNA half-tubes to cover one side of an encapsulated cargo, followed by forming full nanotubes to complete the encapsulation. The structures are assembled from six- and eight-helix bundles, with a typical six helix-based tube composed of two different TX molecules with a bend of 120° between their two DX components. Two molecules of the resulting motif join face-to-face to produce a tube. The eight-helix bundle derives from a single arched four-helix, which links with another copy of itself to give rise to the eight-helix tube. Such half-tube components formed nanotube arrays of regular thicknesses.

Based on this example, most designed tubes based on crossover motifs could be assigned to two main design categories. In one, nanotubes are built from monomers that polymerize with their terminal sticky ends under a non-planar angle leading to 2D arrays, which give tubes by closing on themselves. This type is prone to variations in size and shape due to axial rotation, leading to slanted or skewed nanotubes. In the other, the tube formation is strongly correlated with the exact number of helices in a building block with a strict requirement for a bend angle between crossover components. The latter is generated by strand switching, which in principle can be designed at specific positions to lead to specific structures. Both cases, however, permit a degree of promiscuity, which is probably due to the multistrand nature of the DX motifs, prompting the question of how many strands are required to produce nanotubes. In part, this question can be answered by designing nanotubes assembled from a minimum number of strands, which necessitates sequence symmetry in building a DX motif from a single strand.

The strand was designed to have four palindromic (self-complementary) segments. These associate with other copies of themselves, forming a two-stranded complex. The motif contains two duplex domains and four single-stranded overhangs, thereby assembling into 2D arrays. Because the following folding of the arrays into tubes is an intracomplexation process, as opposed to intercomplex growth of extended 2D lattices, nanotube assemblies are kinetically more favourable at low DNA concentrations. Low concentrations could also help avoid the formation of amorphous structures, preventing the palindromes from associating into complexes other than the target. An exclusive morphology represented by long nanotubes uniform in height and width confirmed these assumptions.[172]

Nanotubes are characteristically rigid structures. The possibility of organizing them into specific arrays presents a certain technological interest, but requires nanotubes to be periodically aligned, which is hampered by their rigidity. An effective strategy to align nanotubes into well-defined arrays is to combine top-down and bottom-up methods. Nanotubes can be assembled by annealing from a pool of single-stranded DNA, followed by their random deposition on solid substrates and alignment into arrays by a surface-patterning technique, *e.g.* soft lithography. This allows the versatile manipulation of the nanotubes within the limits of their morphological integrity.[173]

2.5.3 Biofunctional 3D DNA Origami

However, an ultimate advantage of DNA is its ability to self-code its assembly into discrete nanoscale forms of any description. DNA origami designs provide ample evidence for this. Starting with the first examples of 3D origami,[174,175] successfully demonstrating the suitability of DNA for the construction of various geometric shapes, DNA is readily adapted for dynamic 3D structures designed with a specific function in mind. Given the natural role of DNA to reproduce life, it was long thought about attempting to reintroduce DNA in the form of nanostructures into the cell. Nature employs specialized protein materials, such as viruses, to encapsulate and deliver nucleic acids. With its negative charge, DNA needs an auxiliary means for transport, which viral proteins provide. Indeed, DNA origami structures coated with viral proteins demonstrate enhanced cellular uptake.[176]

Viruses are also geometric symmetry-driven cages the structure of which can be readily emulated by DNA designs. Cage- or cube-like structures can be designed using origami techniques.[175,177,178] Three-dimensional

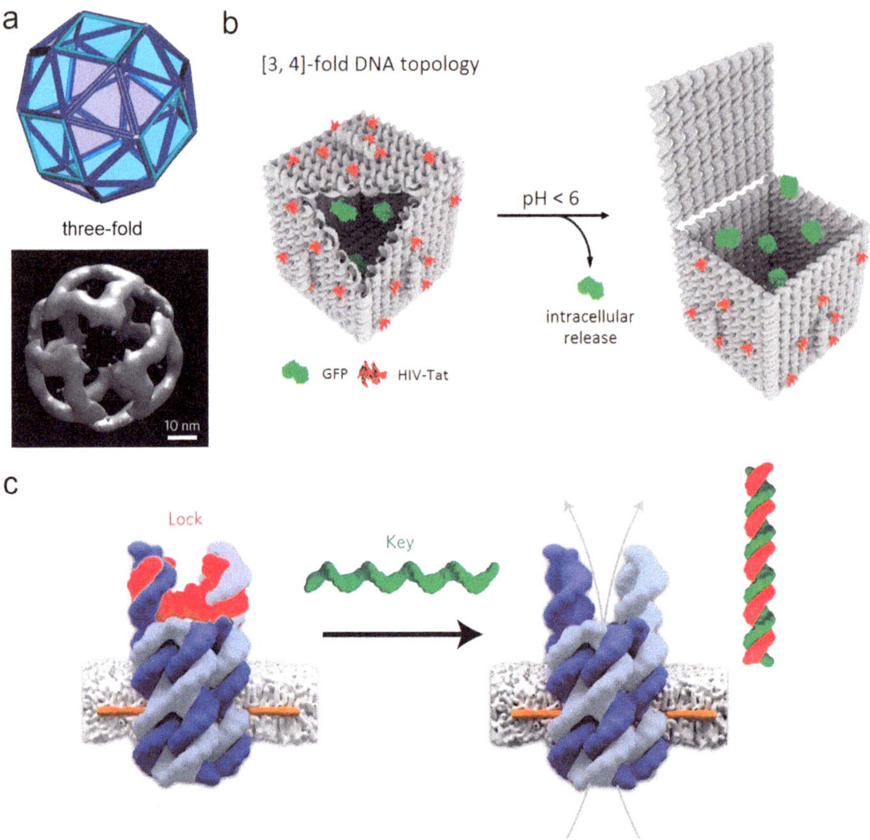

Figure 2.20 3D DNA origami. (a) A DNA snub cube design (top) with density maps reconstructed from cryo-electron microscopy images (bottom) along the threefold rotational symmetry axis. Reproduced from ref. 177 with permission from Springer Nature, Copyright 2015. (b) DNA origami everted virus: a self-assembled DNA box encapsulating green fluorescent protein (GFP) (green) and decorated with HIV-Tat cell-penetrating motifs facilitates the intracellular uptake of the box. Once in the cell, the lid of the box opens up in response to changes in pH, releasing the cargo. Reproduced from ref. 179 with permission from American Chemical Society, Copyright 2018. (c) DNA nanopore assembled from six DNA strands alternately in dark and pale blue. The pore is modified with cholesterol-based anchors (orange), facilitating the incorporation of the nanopore into phospholipid bilayers (grey). The pore is closed (left). When its lock DNA strand (red) hybridizes to key DNA (green), the pore opens. Reproduced from ref. 183 with permission from Springer Nature, Copyright 2016.

wireframe scaffolds assembled with multi-arm junction vertices produce discrete geometries, which can serve as an approximation of cage-like structures, Figure 2.20a.

If made reversible, these forms can provide unique functions such as controlled release of cargo in response to changes in external stimuli.[178]

Since viruses constitute protein cages encapsulating genetic material, DNA cage topologies can be used as everted viruses encapsulating proteins. For instance, a DNA box designed with a controllable lid[178] can be decorated with transduction peptide domains promoting the intracellular uptake of the box.[179] The box itself has three- and fourfold rotational symmetry, thus termed [3,4]-fold DNA topology, and provides an interior space significant enough to encapsulate large proteins, such as green fluorescent protein, Figure 2.20b. Once in the cell, the lid of the box opens up, releasing protein molecules into the cytoplasm, which can be monitored by flow cytometry and live cell imaging. In the harsh cellular environments, the box may not need a lid as it may disassemble faster.[179] For *in vivo* applications, the requirements for stability of the DNA structures are more significant, prompting the need for robust protection. An effective solution can be found in the encapsulation of DNA origami in a lipid coat in a manner similar to that of enveloped viruses.[180]

A similar logic of adapting DNA designs for protein-specialized functions has been applied to the design of nanopores spanning cell membranes.[181–183] Naturally occurring pores are assembled from hydrophobic and cationic pore-forming proteins or peptides, which can incorporate into the hydrophobic, anionic phospholipid bilayers. DNA is anionic and hydrophilic. Therefore, to anchor and remain in the bilayers, it needs to be modified with hydrophobic moieties, such as cholesterol, Figure 2.20c.[183] Once incorporated, such a DNA nanopore can serve as a transmembrane channel for the transport of charged molecular cargo. The transport is controlled by a hybridization event between lock-and-key DNA strands, regulating the open–closed states of the channel, Figure 2.20c. The capacity of DNA nanopores can be extended to single-molecule protein sensing, including multiplexing.[184] Similar and other 3D designs continue to emerge on a regular basis and are well documented in the literature, which is already too vast to be covered here. To this end, the DNA origami approach has matured to the level of standard design protocols and operating procedures, making the technique accessible to experienced researchers and students alike.[185]

2.6 Outlook

As originally postulated by Seeman,[8] DNA is an effective tool for the design of nanoscale structures. Since then, virtually every geometric form of DNA has been tried. Much research remains to be carried out in the realm of designing nanostructures with increased complexity, aiming to revise the limitations of DNA as a universal nanodesign module. Encouragingly, increasing emphasis is also being placed on functional DNA forms, with careful attempts at assigning specific functions. Promising steps are being made towards atypical applications of DNA assemblies, although one can argue that any application that does not fall within the remit of DNA as a hereditary material is atypical. Nonetheless, notable efforts are being made beyond the use of DNA as a decorative template for site-specific molecular recruitment. DNA nanodesigns as vehicles for intracellular delivery of molecular cargo, matrices supporting cell growth and proliferation, and nanopores spanning phospholipid membranes are classic examples of applying an old material to build old forms with new functions. Future developments will only strengthen this position of DNA in the design of biological nanoscale systems, expanding to the areas of single-molecule imaging, traceable synthetic biology, nanoscale enzymatic and biochemical reactors, and therapeutics.

References

1. J. D. Watson and F. H. C. Crick, Molecular Structure of Nucleic Acids: A Structure for Deoxyribose Nucleic Acid, *Nature*, 1953, **171**, 737–738.
2. M. H. F. Wilkins, A. R. Stokes and H. R. Wilson, Molecular Structure of Nucleic Acids: Molecular Structure of Deoxypentose Nucleic Acids, *Nature*, 1953, **171**, 738–740.
3. J. D. Watson and F. H. C. Crick, Genetical Implications of the Structure of Deoxyribonucleic Acid, *Nature*, 1953, **171**, 964–967.
4. R. E. Franklin and R. G. Gosling, Molecular Configuration in Sodium Thymonucleate, *Nature*, 1953, **171**, 740–741.
5. R. E. Franklin and R. G. Gosling, Evidence for 2-Chain Helix in Crystalline Structure of Sodium Deoxyribonucleate, *Nature*, 1953, **172**, 156–157.
6. N. C. Seeman, DNA in a material world, *Nature*, 2003, **421**, 427–431.
7. Y. J. Chen, B. Groves, R. A. Muscat and G. Seelig, DNA nanotechnology from the test tube to the cell, *Nat. Nanotechnol.*, 2015, **10**, 748–760.
8. N. C. Seeman, Nucleic Acid Junctions and Lattices, *J. Theor. Biol.*, 1982, **99**, 237–247.
9. N. C. Seeman, Macromolecular design, nucleic acid junctions, and crystal formation, *J. Biomol. Struct. Dyn.*, 1985, **3**, 11–33.
10. N. C. Seeman, An overview of structural DNA nanotechnology, *Mol. Biotechnol.*, 2007, **37**, 246–257.
11. F. Hong, F. Zhang, Y. Liu and H. Yan, DNA Origami: Scaffolds for Creating Higher Order Structures, *Chem. Rev.*, 2017, **117**, 12584–12640.
12. E. Winfree, On the computational power of DNA annealing and ligation, *DIMACS: Series in Discrete Mathematics and Theoretical Computer Science*, 1996, vol. 27, pp. 199–219.
13. P. W. K. Rothemund, Folding DNA to create nanoscale shapes and patterns, *Nature*, 2006, **440**, 297–302.
14. F. H. C. Crick, L. Barnett, S. Brenner and R. J. Watts-Tobin, General nature of the genetic code for proteins, *Nature*, 1961, **192**, 1227–1232.
15. F. H. C. Crick, The origin of the genetic code, *J. Mol. Biol.*, 1968, **38**, 367–379.
16. S. Itzkovitz and U. Alon, The genetic code is nearly optimal for allowing additional information within protein-coding sequences, *Genome Res.*, 2007, **17**, 405–412.

17. W. Saenger, *Principles of Nucleic Acid Structure*, Springer Verlag, New York, 1988.
18. E. S. Balakirev and F. J. Ayala, Pseudogenes: are they "junk" or functional DNA?, *Annu. Rev. Genet.*, 2003, **37**, 123–151.
19. E. F. Vanin, Processed Pseudogenes: Characteristics and Evolution, *Annu. Rev. Genet.*, 1985, **19**, 253–272.
20. P. M. Harrison, *et al.*, Molecular Fossils in the Human Genome: Identification and Analysis of the Pseudogenes in Chromosomes 21 and 22, *Genome Res.*, 2002, **12**, 272–280.
21. S. Hirotsune, *et al.*, An expressed pseudogene regulates the messenger-RNA stability of its homologous coding gene, *Nature*, 2003, **423**, 91–96.
22. T. Akera, L. Chmátal, E. Trimm, K. Yang, C. Aonbangkhen, D. M. Chenoweth, C. Janke, R. M. Schultz and M. A. Lampson, Spindle asymmetry drives non-Mendelian chromosome segregation, *Science*, 2017, **358**, 668–672.
23. A. H. J. Wang, *et al.*, Molecular structure of a left-handed double helical DNA fragment at atomic resolution, *Nature*, 1979, **282**, 680–686.
24. A. Rich, A. Nordheim and A. H.-J. Wang, The chemistry and biology of left-handed Z-DNA, *Annu. Rev. Biochem.*, 1984, **53**, 791–846.
25. Q. Du, C. Smith, N. Shiffeldrim, M. Vologodskaia and A. Vologodskii, Cyclization of short DNA fragments and bending fluctuations of the double helix, *Proc. Natl. Acad. Sci. U. S. A.*, 2005, **102**, 5397–5402.
26. S. Cocco, R. Monasson and J. F. Marko, Force and kinetic barriers to initiation of DNA unzipping, *Phys. Rev. E*, 2002, **65**, 041907.
27. I. M. Kulic and H. Schiessel, DNA Spools under Tension, *Phys. Rev. Lett.*, 2004, **92**, 228101.
28. J. Bednar, *et al.*, Determination of DNA Persistence Length by Cryo-electron Microscopy. Separation of the Static and Dynamic Contributions to the Apparent Persistence Length of DNA, *J. Mol. Biol.*, 1995, **254**, 579–594.
29. P. J. Hagerman, Flexibility of DNA, *Annu. Rev. Biophys. Biophys. Chem.*, 1988, **17**, 265–286.
30. V. V. Demidov and M. D. Frank-Kamenetskii, Two sides of the coin: affinity and specificity of nucleic acid interactions, *Trends Biochem. Sci.*, 2004, **29**, 62–71.
31. N. C. Seeman, De Novo Design of Sequences for Nucleic Acid Structural Engineering, *J. Biomol. Struct. Dyn.*, 1990, **8**, 573–581.
32. Y. Yang, T. P. Westcott, S. C. Pedersen, I. Tobias and W. K. Olson, Effects of localized bending on DNA supercoiling, *Trends Biochem. Sci.*, 1995, **20**, 313–319.
33. A. V. Vologodskii, A. V. Lukashin, V. V. Anshelevich and M. D. Frank-Kamenetskii, Fluctuations in superhelical DNA, *Nucleic Acids Res.*, 1979, **6**, 967–982.
34. G. A. L. Chirico, J. Brownian dynamics simulations of supercoiled DNA with bent sequences, *Biophys. J.*, 1996, **71**, 955–971.

35. A. V. Vologodskii and N. R. Cozzarelli, Conformational and Thermodynamic Properties of Supercoiled DNA, *Annu. Rev. Biophys. Biomol. Struct.*, 1994, **23**, 609–643.
36. N. C. Seeman, The Use of Branched DNA for Nanoscale Fabrication, *Nanotechnology*, 1991, **2**, 149–159.
37. N. C. Seeman, Construction of three-dimensional stick figures from branched DNA, *DNA Cell Biol.*, 1991, **10**, 475–486.
38. N. C. Seeman, The design of single-stranded nucleic acid knots, *Mol. Eng.*, 1992, **2**, 297–307.
39. N. C. Seeman, DNA engineering and its application to nanotechnology, *Trends Biotechnol.*, 1999, **17**, 437–443.
40. N. C. Seeman, *et al.*, New motifs in DNA nanotechnology, *Nanotechnology*, 1998, **9**, 257–273.
41. Y. Zhang and N. C. Seeman, Construction of a DNA-Truncated Octahedron, *J. Am. Chem. Soc.*, 1994, **116**, 1661–1669.
42. C. Mao, W. Sun and N. C. Seeman, Assembly of Borromean rings from DNA, *Nature*, 1997, **386**, 137–138.
43. C. Liang and K. Mislow, On Borromean links, *J. Math. Chem.*, 1994, **16**, 27–35.
44. X. Li, X. Yang, J. Qi and N. C. Seeman, Antiparallel DNA double crossover molecules as components for nanoconstruction, *J. Am. Chem. Soc.*, 1996, **118**, 6131–6140.
45. T.-J. Fu and N. C. Seeman, DNA double crossover structures, *Biochemistry*, 1993, **32**, 3211–3220.
46. T. H. LaBean, *et al.*, Construction, Analysis, Ligation, and Self-Assembly of DNA Triple Crossover Complexes, *J. Am. Chem. Soc.*, 2000, **122**, 1848–1860.
47. H. Yan and N. C. Seeman, Edge-sharing motifs in structural DNA nanotechnology, *J. Supramol. Chem.*, 2001, **1**, 229–237.
48. X. Zhang, H. Yan, Z. Shen and N. C. Seeman, Paranemic Cohesion of Topologically-Closed DNA Molecules, *J. Am. Chem. Soc.*, 2002, **124**, 12940–12941.
49. U. Feldkamp, R. Wacker, W. Banzhaf and C. M. Niemeyer, Microarray-Based *in vitro* Evaluation of DNA Oligomer Libraries Designed *in silico*, *ChemPhysChem*, 2004, **5**, 367–372.
50. M. Zuker, Mfold web server for nucleic acid folding and hybridization prediction, *Nucleic Acids Res.*, 2003, **31**, 3406–3415.
51. N. R. Markham and M. Zuker, DINAMelt web server for nucleic acid melting prediction, *Nucleic Acids Res.*, 2005, **33**, W577–W581.
52. N. Le Novere, MELTING, computing the melting temperature of nucleic acid duplex, *Bioinformatics*, 2001, **17**, 1226–1227.
53. K. J. Breslauer, R. Frank, H. Blocker and L. A. Marky, Predicting DNA duplex stability from the base sequence, *Proc. Natl. Acad. Sci. U. S. A.*, 1986, **83**, 3746–3750.
54. M. R. Shortreed, S. B. Chang, D. Hong, M. Phillips, B. Campion, D. C. Tulpan, A. Condon, H. H. Hoos and L. M. Smith, A thermodynamic approach to designing structure-free combinatorial DNA word sets, *Nucleic Acids Res.*, 2005, **22**, 4965–4977.

55. M. R. Shortreed, *et al.*, A thermodynamic approach to designing structure-fr4ee combinatorial DNA word sets, *Nucleic Acids Res.*, 2005, **33**, 4965–4977.
56. R. M. Dirks, M. Lin, E. Winfree and N. A. Pierce, Paradigms for computational nucleic acid design, *Nucleic Acids Res.*, 2004, **32**, 1392–1403.
57. J. Sager and D. Stefanovic, in *DNA Computing*, 2006, pp. 275–289.
58. I. L. Hofacker, Vienna RNA secondary structure server, *Nucleic Acids Res.*, 2003, **31**, 3429–3431.
59. J. H. Reif, T. H. LaBean, S. Sahu, H. Yan and P. Yin, in *Unconventional Programming Paradigms*, 2005, pp. 173–187.
60. A. A. Podtelezhnikov, N. R. Cozzarelli and A. V. Vologodskii, Equilibrium distributions of topological states in circular DNA: Interplay of supercoiling and knotting, *Proc. Natl. Acad. Sci. U. S. A.*, 1999, **96**, 12974–12979.
61. A. A. Podtelezhnikov, C. Mao, N. C. Seeman and A. Vologodskii, Multimerization-Cyclization of DNA Fragments as a Method of Conformational Analysis, *Biophys. J.*, 2000, **79**, 2692–2704.
62. A. Vologodskii, Exploiting circular DNA, *Proc. Natl. Acad. Sci. U. S. A.*, 1998, **95**, 4092–4093.
63. Q. Du, A. Kotlyar and A. Vologodskii, Kinking the double helix by bending deformation, *Nucleic Acids Res.*, 2008, **36**, 1120–1128.
64. H. Jacobson and W. H. Stockmayer, Intramolecular Reaction in Polycondensations. I. The Theory of Linear Systems, *J. Chem. Phys.*, 1950, **18**, 1600–1606.
65. D. Shore, J. Langowski and R. L. Baldwin, DNA flexibility studied by covalent closure of short fragments into circles, *Proc. Natl. Acad. Sci. U. S. A.*, 1981, **78**, 4833–4837.
66. D. Shore, J. Langowski and R. L. Baldwin, DNA flexibility studied by covalent closure of short fragments into circles, *Proc. Natl. Acad. Sci. U. S. A.*, 1981, **78**, 4833–4837.
67. P. Sa-Ardyen, A. V. Vologodskii and N. C. Seeman, The Flexibility of DNA Double Crossover Molecules, *Biophys. J.*, 2003, **84**, 3829–3837.
68. J. SantaLucia and D. Hicks, The thermodynamics of DNA structural motifs, *Annu. Rev. Biophys. Biomol. Struct.*, 2004, **33**, 415–440.
69. A. Brenneman and A. Condon, Strand design for bio-molecular computation, *Theor. Comput. Sci.*, 2002, **287**, 39.
70. L. M. Adleman, Molecular computation of solutions to combinatorial problems, *Science*, 1994, **266**, 1021–1024.
71. K. J. Baeyens, H. L. De Bondt, A. Pardi and S. R. Holbrook, A curved RNA helix incorporating an internal loop with GA and AA non-Watson-Crick base pairing, *Proc. Natl. Acad. Sci. U. S. A.*, 1996, **93**, 12851–12855.
72. S.-H. Ke and R. M. Wartell, Influence of Neighboring Base Pairs on the Stability of Single Base Bulges and Base Pairs in a DNA Fragment, *Biochemistry*, 1995, **34**, 4593–4600.
73. I. L. Kuznetsova, M. A. Zenkova, H. J. Gross and V. V. Vlassov, Enhanced RNA cleavage within bulge-loops by an artificial ribonuclease, *Nucleic Acids Res.*, 2005, **33**, 1201–1212.

74. J.-Y. Wang and K. Drlica, Modeling hybridization kinetics, *Math. Biosci.*, 2003, **183**, 37–47.
75. A.-C. Déclais and D. M. J. Lilley, New insight into the recognition of branched DNA structure by junction-resolving enzymes, *Curr. Opin. Struct. Biol.*, 2008, **18**, 86–95.
76. N. D. F. Grindley, K. L. Whiteson and P. A. Rice, Mechanisms of Site-Specific Recombination, *Annu. Rev. Biochem.*, 2006, **75**, 567–605.
77. R. Holliday, A mechanism for gene conversion in fungi, *Genet. Res.*, 1964, **5**, 282–304.
78. Y. Wang, J. E. Mueller, B. Kemper and N. C. Seeman, Assembly and characterization of five-arm and six-arm DNA branched junctions, *Biochemistry*, 1991, **30**, 5667–5674.
79. X. Wang and N. C. Seeman, Assembly and Characterization of 8-Arm and 12-Arm DNA Branched Junctions, *J. Am. Chem. Soc.*, 2007, **129**, 8169–8176.
80. B. Liu, N. B. Leontis and N. C. Seeman, Bulged 3-arm DNA branched junctions as components for nanoconstruction, *Nanobiology*, 1994, **3**, 177–188.
81. J. Qi, X. Li, X. Yang and N. C. Seeman, The ligation of triangles built from bulged three-arm DNA branched junctions, *J. Am. Chem. Soc.*, 1996, **118**, 6121–6130.
82. J. Chen, N. R. Kallenbach and N. C. Seeman, A Specific Quadrilateral Synthesized from DNA Branched Junctions, *J. Am. Chem. Soc.*, 1989, **111**, 6402–6407.
83. J. Chen and N. C. Seeman, Synthesis from DNA of a molecule with the connectivity of a cube, *Nature*, 1991, **350**, 631–633.
84. J. E. Mueller, S. M. Du and N. C. Seeman, The design and synthesis of a knot from single-stranded DNA, *J. Am. Chem. Soc.*, 1991, **113**, 6306–6308.
85. N. C. Seeman, DNA Nicks and Nodes and Nanotechnology, *Nano Lett.*, 2001, **1**, 22–26.
86. S. M. Du, S. Zhang and N. C. Seeman, DNA junctions, antijunctions, and mesojunctions, *Biochemistry*, 1992, **31**, 10955–10963.
87. H. Wang and N. C. Seeman, Structural Domains of DNA Mesojunctions, *Biochemistry*, 1995, **34**, 920–929.
88. M. E. Churchill, T. D. Tullius, N. R. Kallenbach and N. C. Seeman, A Holliday recombination intermediate is twofold symmetric, *Proc. Natl. Acad. Sci. U. S. A.*, 1988, **85**, 4653–4656.
89. S. M. Du, B. D. Stollar and N. C. Seeman, A synthetic DNA molecule in three knotted topologies, *J. Am. Chem. Soc.*, 1995, **117**, 1194–1200.
90. N. C. Seeman, S. Zhang and J. Chen, DNA Nanoconstructions, *J. Vac. Sci. Technol., A*, 1994, **12**, 1895–1903.
91. Y. Li, *et al.*, Controlled assembly of dendrimer-like DNA, *Nat. Mater.*, 2004, **3**, 38–42.
92. T. F. Roth and D. R. Helinski, Evidence for circular DNA forms of a bacterial plasmid, *Proc. Natl. Acad. Sci. U. S. A.*, 1967, **58**, 650–657.

93. T. Cremer and C. Cremer, Chromosome territories, nuclear architecture and gene regulation in mammalian cells, *Nat. Rev. Genet.*, 2001, **2**, 292–301.
94. K. Timmis, F. Cabello and S. N. Cohen, Covalently closed circular DNA molecules of low superhelix density as intermediate forms in plasmid replication, *Nature*, 1976, **261**, 512–516.
95. J. Vinograd, J. Lebowitz, R. Radloff, R. Watson and P. Laipis, The twisted circular form of polyoma viral DNA, *Proc. Natl. Acad. Sci. U. S. A.*, 1965, **53**, 1104–1111.
96. S. Y. Shaw and J. C. Wang, Knotting of a DNA chain during ring closure, *Science*, 1993, **260**, 533–536.
97. A. V. Vologodskii and N. R. Cozzarelli, Effect of supercoiling on the juxtaposition and relative orientation of DNA sites, *Biophys. J.*, 1996, **70**, 2548–2556.
98. A. V. Vologodskii and N. R. Cozzarelli, Supercoiling, knotting, looping, and other large-scale conformational properties of DNA, *Curr. Opin. Struct. Biol.*, 1994, **4**, 372–375.
99. F. B. Fuller, The Writhing Number of a Space Curve, *Proc. Natl. Acad. Sci. U. S. A.*, 1971, **68**, 815–819.
100. J. H. White, Self-Linking and the Gauss Integral in Higher Dimensions, *Am. J. Math.*, 1969, **91**, 693–728.
101. S. A. Wasserman and N. R. Cozzarelli, Biochemical topology: Applications to DNA recombination and replication, *Science*, 1986, **232**, 951–960.
102. C. C. Adams, *The Knot Book: An Elementary Introduction to the Mathematical Theory of Knots*, W. H. Freeman & Company, New York, 2004.
103. A. V. Vologodskii, *et al.*, Sedimentation and electrophoretic migration of DNA knots and catenanes, *J. Mol. Biol.*, 1998, **278**, 1–3.
104. J. Hoste, M. Thistlethwaite and J. Weeks, The first 1701936 knots, *Math. Intell.*, 1998, **20**, 33–48.
105. S. A. Wasserman, J. M. Dungan and N. R. Cozzarelli, Discovery of a predicted DNA knot substantiates a model for site-specific recombination, *Science*, 1985, **229**, 171–174.
106. S. A. Wasserman and N. R. Cozzarelli, Supercoiled DNA-directed knotting by T4 topoisomerase, *J. Biol. Chem.*, 1991, **266**, 20567–20573.
107. P. O. Brown and N. R. Cozzarelli, A sign inversion mechanism for enzymatic supercoiling of DNA, *Science*, 1979, **206**, 1081–1083.
108. V. V. Rybenkov, C. Ullsperger, A. V. Vologodskii and N. R. Cozzarelli, Simplification of DNA topology below equilibrium values by type II topoisomerases, *Science*, 1997, **277**, 690–693.
109. F. B. Dean, A. Stasiak, T. Koller and N. R. Cozzarelli, Duplex DNA knots produced by Escherichia coli topoisomerase I. Structure and requirements for formation, *J. Biol. Chem.*, 1985, **260**, 4975–4983.
110. L. F. Liu, C. C. Liu and B. M. Alberts, Type II DNA topoisomerases: enzymes that can unknot a topologically knotted DNA molecule *via* a reversible double-strand break, *Cell*, 1980, 697–707.

111. S. M. Du, H. Wang, Y.-C. Tse-Dinh and N. C. Seeman, Topological transformations of synthetic DNA knots, *Biochemistry*, 1995, **34**, 673–682.
112. H. Wang, R. J. Di Gate and N. C. Seeman, An RNA topoisomerase, *Proc. Natl. Acad. Sci. U. S. A.*, 1996, **93**, 9477–9482.
113. A. Xayaphoummine, T. Bucher and H. Isambert, Kinefold web server for RNA/DNA folding path and structure prediction including pseudoknots and knots, *Nucleic Acids Res.*, 2005, **33**, W605–W610.
114. Y. P. Ohayon, R. Sha, O. Flint, A. R. Chandrasekaran, H. O. Abdallah, T. Wang, X. Wang, X. Zhang and N. C. Seeman, Topological Linkage of DNA Tiles Bonded by Paranemic Cohesion, *ACS Nano*, 2015, **9**, 10296–10303.
115. S. D. Levene, C. Donahue, T. C. Boles and N. R. Cozzarelli, Analysis of the structure of dimeric DNA catenanes by electron microscopy, *Biophys. J.*, 1995, **69**, 1036–1045.
116. A. Bucka and A. Stasiak, Construction and electrophoretic migration of single-stranded DNA knots and catenanes, *Nucleic Acids Res.*, 2002, **30**, e24.
117. D. E. Adams, E. M. Shekhtman, E. L. Zechiedrich, M. B. Scmid and N. R. Cozzarelli, The role of topoisomerase IV in partitioning bacterial replicons and the structure of catenated intermediates in DNA repliation, *Cell*, 1992, **71**, 277–288.
118. J. B. Bliska and N. R. Cozzarelli, Use of site-specific recombination as a probe of DNA structure and metabolism *in vivo*, *J. Mol. Biol.*, 1987, **194**, 205–218.
119. T.-J. Fu, Y. C. Tse-Dinh and N. C. Seeman, Holliday junction crossover topology, *J. Mol. Biol.*, 1994, **236**, 91–105.
120. J. Chen and N. C. Seeman, The electrophoretic properties of a DNA cube and its substructure catenanes, *Electrophoresis*, 1991, **12**, 607–611.
121. S. J. Cantrill, K. S. Chichak, A. J. Peters and J. F. Stoddart, Nanoscale Borromean Rings, *Acc. Chem. Res.*, 2005, **38**, 1–9.
122. H. Qiu, J. C. Dewan and N. C. Seeman, A DNA decamer with a sticky end: the crystal structure of d-CGACGATCGT, *J. Mol. Biol.*, 1997, **267**, 881–898.
123. T.-J. Fu, Y.-C. Tse-Dinh and N. C. Seeman, Holliday junction crossover topology, *J. Mol. Biol.*, 1994, **236**, 91–105.
124. E. Winfree, F. Liu, L. A. Wenzler and N. C. Seeman, Design and self-assembly of two-dimensional DNA crystals, *Nature*, 1998, **394**, 539–544.
125. N. B. Leontis, *et al.*, A model for the solution structure of a branched, three-strand DNA complex, *J. Biomol. Struct. Dyn.*, 1993, **11**, 215–223.
126. X. Yang, L. A. Wenzler, J. Qi, X. Li and N. C. Seeman, Ligation of DNA Triangles Containing Double Crossover Molecules, *J. Am. Chem. Soc.*, 1998, **120**, 9779–9786.
127. B. Ding, R. Sha and N. C. Seeman, Pseudohexagonal 2D DNA Crystals from Double Crossover Cohesion, *J. Am. Chem. Soc.*, 2004, **126**, 10230–10231.
128. R.-I. Ma, N. R. Kallenbach, R. D. Sheardy, M. L. Petrillo and N. C. Seeman, Three-arm nucleic acid junctions are flexible, *Nucleic Acids Res.*, 1986, **14**, 9745–9753.

129. D. Liu, M. Wang, Z. Deng, R. Walulu and C. Mao, Tensegrity: Construction of Rigid DNA Triangles with Flexible Four-Arm DNA Junctions, *J. Am. Chem. Soc.*, 2004, **126**, 2324–2325.
130. F. Mathieu, *et al.*, Six-Helix Bundles Designed from DNA, *Nano Lett.*, 2005, **5**, 661–665.
131. P. E. Constantinou, *et al.*, Double cohesion in structural DNA nanotechnology, *Org. Biomol. Chem.*, 2006, **4**, 3414–3419.
132. J. C. Wang, Helical Repeat of DNA in Solution, *Proc. Natl. Acad. Sci. U. S. A.*, 1979, **76**, 200–203.
133. D. Rhodes and A. Klug, Helical periodicity of DNA determined by enzyme digestion, *Nature*, 1980, **286**, 573–578.
134. R. Sha, H. Iwasaki, F. Liu, H. Shinagawa and N. C. Seeman, Cleavage of Symmetric Immobile DNA Junctions by Escherichia coli RuvC, *Biochemistry*, 2000, **39**, 11982–11988.
135. R. Sha, F. Liu and N. C. Seeman, Direct Evidence for Spontaneous Branch Migration in Antiparallel DNA Holliday Junctions, *Biochemistry*, 2000, **39**, 11514–11522.
136. N. C. Seeman, *et al.*, Chemical Synthesis of Nanostructures, *Mater. Res. Soc. Symp. Proc.*, 1994, **330**, 45–56.
137. Z. Shen, H. Yan, T. Wang and N. C. Seeman, Paranemic Crossover DNA: A Generalized Holliday Structure with Applications in Nanotechnology, *J. Am. Chem. Soc.*, 2004, **126**, 1666–1674.
138. H. Yan, S. H. Park, G. Finkelstein, J. H. Reif and T. H. LaBean, DNA-Templated Self-Assembly of Protein Arrays and Highly Conductive Nanowires, *Science*, 2003, **301**, 1882–1884.
139. W. Liu, X. Wang, T. Wang, R. Sha and N. C. Seeman, PX DNA Triangle Oligomerized Using a Novel Three-Domain Motif, *Nano Lett.*, 2008, **8**, 317–322.
140. H. Yan, X. Zhang, Z. Shen and N. C. Seeman, A robust DNA mechanical device controlled by hybridization topology, *Nature*, 2002, **415**, 62–65.
141. S. Liao and N. C. Seeman, Translation of DNA Signals into Polymer Assembly Instructions, *Science*, 2004, **306**, 2072–2074.
142. B. Ding and N. C. Seeman, Operation of a DNA Robot Arm Inserted into a 2D DNA Crystalline Substrate, *Science*, 2006, **314**, 1583–1585.
143. A. V. Garibotti, S. Liao and N. C. Seeman, A Simple DNA-Based Translation System, *Nano Lett.*, 2007, **7**, 480–483.
144. B. H. Robinson and N. C. Seeman, The design of a biochip: a self-assembling molecular-scale memory device, *Prot. Eng.*, 1987, **1**, 295–300.
145. C. Mao, W. Sun, Z. Shen and N. C. Seeman, A nanomechanical device based on the B-Z transition of DNA, *Nature*, 1999, **397**, 144–146.
146. C. Lin, X. Wang, Y. Liu, N. C. Seeman and H. Yan, Rolling Circle Enzymatic Replication of a Complex Multi-Crossover DNA Nanostructure, *J. Am. Chem. Soc.*, 2007, **129**, 14475–14481.
147. C. Mao, W. Sun and N. C. Seeman, Designed Two-Dimensional DNA Holliday Junction Arrays Visualized by Atomic Force Microscopy, *J. Am. Chem. Soc.*, 1999, **121**, 5437–5443.

148. R. Sha, F. Liu, D. P. Millar and N. C. Seeman, Atomic force microscopy of parallel DNA branched junction arrays, *Chem. Biol.*, 2000, **7**, 743–751.
149. S. H. Park, G. Finkelstein and T. H. LaBean, Stepwise self-assembly of DNA tile lattices using dsDNA bridges, *J. Am. Chem. Soc.*, 2008, **130**, 40–41.
150. J. Malo, J. C. Mitchell, C. Vénien-Bryan, J. R. Harris, H. Wille, D. J. Sherratt and A. J. Turberfield, Engineering a 2D protein-DNA crystal, *Angew. Chem., Int. Ed.*, 2005, **44**, 3057–3061.
151. X. Qi, F. Zhang, Z. Su, S. Jiang, D. Han, B. Ding, Y. Liu, W. Chiu, P. Yin and H. Yan, Programming molecular topologies from single-stranded nucleic acids, *Nat. Commun.*, 2018, **9**, 4579.
152. W. M. Shih, J. D. Quispe and G. F. Joyce, A 1.7-kilobase single-stranded DNA that folds into a nanoscale octahedron, *Nature*, 2004, **427**, 618–621.
153. R. P. Goodman, *et al.*, Rapid Chiral Assembly of Rigid DNA Building Blocks for Molecular Nanofabrication, *Science*, 2005, **310**, 1661–1665.
154. R. P. Goodman, R. M. Berry and A. J. Turberfield, The single-step synthesis of a DNA tetrahedron, *Chem. Commun.*, 2004, 1372–1373.
155. C. M. Erben, R. P. Goodman and A. J. Turberfield, A Self-Assembled DNA Bipyramid, *J. Am. Chem. Soc.*, 2007, **129**, 6992–6993.
156. R. P. Goodman, *et al.*, Reconfigurable, braced, three-dimensional DNA nanostructures, *Nat. Nanotechnol.*, 2008, **3**, 93–96.
157. B. Yurke, A. J. Turberfield, A. P. Mills, F. C. Simmel and J. L. Neumann, A DNA-fuelled molecular machine made of DNA, *Nature*, 2000, **406**, 605–608.
158. J. von Neumann, *Theory of Self-reproducing Automata*, ed. Burks, A., University of Illinois Press, Urbana, Illinois, 1966.
159. H. Wang, *Dominoes and the AEA Case of the Decision Problem*, ed. Fox, J., Polytechnic Press, Brooklyn, N.Y, 1963.
160. P. W. K. Rothemund, N. Papadakis and E. Winfree, Algorithmic Self-Assembly of DNA Sierpinski Triangles, *PLoS Biol.*, 2004, **2**, 2041–2053.
161. C. Mao, T. H. LaBean, J. H. Reif and N. C. Seeman, Logical computation using algorithmic self-assembly of DNA triple-crossover molecules, *Nature*, 2000, **407**, 493–496.
162. H. Yan, T. H. LaBean, L. Feng and J. H. Reif, Directed nucleation assembly of DNA tile compelxes for barcode-patterned lattices, *Proc. Natl. Acad. Sci. U. S. A.*, 2003, **100**, 8103–8108.
163. F. Hong, F. Zhang, Y. Liu and H. Yan, DNA Origami: Scaffolds for Creating Higher Order Structures, *Chem. Rev.*, 2017, **117**, 12584–12640.
164. F. A. Aldaye, W. T. Senapedis, P. A. Silver and J. C. Way, A structurally tunable DNA-based extracellular matrix, *J. Am. Chem. Soc.*, 2010, **132**, 14727–14729.
165. G. Grossi, M. Dalgaard Ebbesen Jepsen, J. Kjems and E. S. Andersen, Control of enzyme reactions by a reconfigurable DNA nanovault, *Nat. Commun.*, 2017, **8**, 992.

166. K. Zhou, Y. Zhou, V. Pan, Q. Wang and Y. Ke, Programming Dynamic Assembly of Viral Proteins with DNA Origami, *J. Am. Chem. Soc.*, 2020, **142**, 5929–5932.
167. D. Liu, S. H. Park, J. H. Reif and T. H. LaBean, DNA nanotubes self-assembled from triple-crossover tiles as templates for conductive nanowires, *Proc. Natl. Acad. Sci. U. S. A.*, 2004, **101**, 717–722.
168. W. B. Sherman and N. C. Seeman, Design of Minimally Strained Nucleic Acid Nanotubes, *Biophys. J.*, 2006, **90**, 4546–4557.
169. J. C. Mitchell, J. R. Harris, J. Malo, J. Bath and A. J. Turberfield, Self-Assembly of Chiral DNA Nanotubes, *J. Am. Chem. Soc.*, 2004, **126**, 16342–16343.
170. P. W. K. Rothemund, *et al.*, Design and Characterization of Programmable DNA Nanotubes, *J. Am. Chem. Soc.*, 2004, **126**, 16344–16352.
171. A. Kuzuya, R. Wang, R. Sha and N. C. Seeman, Six-Helix and Eight-Helix DNA Nanotubes Assembled from Half-Tubes, *Nano Lett.*, 2007, **7**, 1757–1763.
172. H. Liu, Y. Chen, Y. He, A. E. Ribbe and C. Mao, Approaching The Limit: Can One DNA Oligonucleotide Assemble into Large Nanostructures?, *Angew. Chem., Int. Ed.*, 2006, **45**, 1942–1945.
173. C. Lin, *et al.*, Functional DNA Nanotube Arrays: Bottom-Up Meets Top-Down, *Angew. Chem., Int. Ed.*, 2007, **46**, 6089–6092.
174. S. M. Douglas, H. Dietz, T. Liedl, B. Högberg, F. Graf and W. M. Shih, Self-assembly of DNA into nanoscale three-dimensional shapes, *Nature*, 2009, **459**, 414–418.
175. E. Benson, A. Mohammed, J. Gardell, S. Masich, E. Czeizler, P. Orponen and B. Högberg, DNA rendering of polyhedral meshes at the nanoscale, *Nature*, 2015, **523**, 441–444.
176. J. Mikkila, A. P. Eskelinen, E. H. Niemela, V. Linko, M. J. Frilander, P. Torma and M. A. Kostiainen, Virus-Encapsulated DNA Origami Nanostructures for Cellular Delivery, *Nano Lett.*, 2014, **14**, 2196–2200.
177. R. Veneziano, S. Ratanalert, K. Zhang, F. Zhang, H. Yan, W. Chiu and M. Bathe, Designer nanoscale DNA assemblies programmed from the top down, *Science*, 2016, **352**, 1534.
178. E. S. Andersen, *et al.*, Self-assembly of a nanoscale DNA box with a controllable lid, *Nature*, 2009, **459**, 73–76.
179. J. R. Burns, B. Lamarre, A. L. B. Pyne, J. E. Noble and M. G. Ryadnov, DNA Origami Inside-Out Viruses, *ACS Synth. Biol.*, 2018, **7**, 767–773.
180. S. D. Perrault and W. M. Shih, Virus-Inspired Membrane Encapsulation of DNA Nanostructures to Achieve *in Vivo* Stability, *ACS Nano*, 2014, **8**, 5132–5140.
181. M. Langecker, V. Arnaut, T. G. Martin, J. List, S. Renner, M. Mayer, H. Dietz and F. C. Simmel, Synthetic lipid membrane channels formed by designed DNA nanostructures, *Science*, 2012, **338**, 932–936.
182. N. A. Bell, C. R. Engst, M. Ablay, G. Divitini, C. Ducati, T. Liedl and U. F. Keyser, DNA origami nanopores, *Nano Lett.*, 2012, **12**, 512–517.

183. J. R. Burns, A. Seifert, N. Fertig and S. Howorka, A biomimetic DNA-based channel for the ligand-controlled transport of charged molecular cargo across a biological membrane, *Nat. Nanotechnol.*, 2016, **11**(2), 152–156.
184. N. A. Bell and U. F. Keyser, Digitally encoded DNA nanostructures for multiplexed, single-molecule protein sensing with nanopores, *Nat. Nanotechnol.*, 2016, **11**, 645–651.
185. S. Dey, *et al.*, DNA origami, *Nat. Rev. Dis. Primers*, 2021, **1**, 13.

CHAPTER 3

Bioinspired Nanodesigns for Molecular Encapsulation

3.1 Introduction

Molecular encapsulation is a universal phenomenon that biological systems employ to support a great variety of functions.[1] Being a stimuli-responsive process, it serves to compartmentalize biochemical reactions and assembly processes in a reversible manner, regulate concentration gradients and enable molecular transport.[2] The versatility of molecular encapsulation extends to individual cells and beyond, impacting on biology across multiple length scales, which renders it a hierarchical process. Despite the diversity of functions it supports, the process unifies all under the same basic principles of sustaining enclosed environments and hosting guest elements at the nanoscale.[3]

Nature assigns discrete systems acting as caged receptors and host–guest carriers capitalizing on the precision of molecular recognition patterns promoting stoichiometric ratios of constituent components.[4] Synthetic analogues of naturally occurring systems and also artificial constructions tend to prefer non-covalent encapsulating designs. Such systems are used to support biochemical conversions and typically involve trapping mechanisms, which allows them to capture and protect reactants.[5] The starting materials and conversion intermediates remain "protected" until the conversion is complete, and products are released. Since non-covalent formats imply reversibility, molecular recognition complements encapsulation, allowing for the design of predictive interactions between the trapping and the entrapped components. Other non-covalent systems act as three-dimensional surroundings to spatially confine guest entities and molecules, without necessarily supporting specific

Nanoscience & Nanotechnology Series No. 53
Bionanodesign: Old Forms for New Functions, 2nd Edition
By Maxim Ryadnov

Published by the Royal Society of Chemistry, www.rsc.org

interactions with them. Designs of this type find use in molecular transport, *e.g.* gene delivery systems or endosomes mediating their transfer. The release of encapsulated guests may not be compulsory for this type, creating support for prolonged storage and complex, sequential biochemical transformations. Non-covalent encapsulants do not require high precision and directionality of guest–host interactions that are important for covalent protecting capsules. The latter typically operate at the level of individual molecules, for which molecular stoichiometry is a decisive factor. To a certain extent it is relevant for non-covalent systems: artificial viruses or virus-like particles encapsulate nucleic acids to a point of stoichiometric saturation with capsid proteins, which is predefined by the way in which the folding of the proteins accommodates the geometry of the capsids. Non-covalent encapsulants appear more versatile and can be applied to a greater number of reactions or guests than covalent host–guest systems. Specificity can be rationally introduced, but there remain generic limitations. As covalent systems, which are normally fixed molecular coordination cages,[6] non-covalent encapsulants self-assemble from spatially defined building blocks, be these protein subunits in capsids or DNA origami, which associate *via* complementary, supramolecular couplings. Failure of such associations can lead to partially assembled or aberrant structures and can promote agglomeration. The hallmark of self-assembly is that it can support hierarchically complex systems accessing nano- to microscopic length scales and higher.[7] This range is beyond the reach of covalent encapsulators since the synthesis of monodisperse and spatially intra-locked molecules of increased complexity poses very costly demands.

Designs capitalizing on the principles of biomolecular self-assembly have grown steadily over the last two decades in diversity and numbers. Different functions are aimed for and different chemistries are being used. The amount of information available exceeds the capacity of a single chapter, but a number of excellent reviews are regularly published on the topic covering specialist subjects, which can give the reader detailed representation of the state-of-the-art and future perspectives. This chapter elaborates on the objective of this book, providing a high-level outline of successes and challenges in the design of nanoscale systems emulating native self-assembling forms, with examples of their use with new or atypical applications. A notable example is the design of artificial antimicrobial viruses that destroy bacteria on contact.

3.2 Old Forms of Macromolecular Encapsulation Inspire Nanoscale Designs

Over the last decade, an earlier focus on supporting material transfer, storage and biochemical conversions has evolved into the design of virus-like structures – capsules, cages, vesicles, particles – which can demonstrate the efficacy and reproducibility of their originators. Thus far, the progress appears disproportionate in that there are more structural than functional mimetics of natural systems. This is understandable: structure defines the biological function of any given assembly, therefore getting it right is a prerequisite for successful designs. Although our understanding of structure–function relationships is improving, it is challenging to produce structurally uniform and discrete encapsulators. Besides, their properties are often predetermined by the physical parameters of their cargo, including size, charge, shape and stability. These are the main factors that impact on the performance and structural consistency of encapsulating systems.

Among all the templates that Nature can offer for encapsulating systems, viruses are perhaps the most attractive, both functionally and aesthetically. The versatility of their assembly inspires artificial designs. The same could be said about other protein cage- or shell-like structures. However, it is the appeal of biological reproducibility which favours viruses as universal design templates.

At its simplest, a virus is a protein cage encapsulating a genetic cargo.[8] As a rule, viruses are monodisperse and stimuli-responsive nanoscale objects that are capable of self-assembly with or without cargo. As archetypal nanomaterials they support a range of sizes, from just a few nanometres to several hundred nanometres, but each virus is remarkably conserved when it comes to its replication, Figure 3.1.[9] Most artificial designs reported to date draw

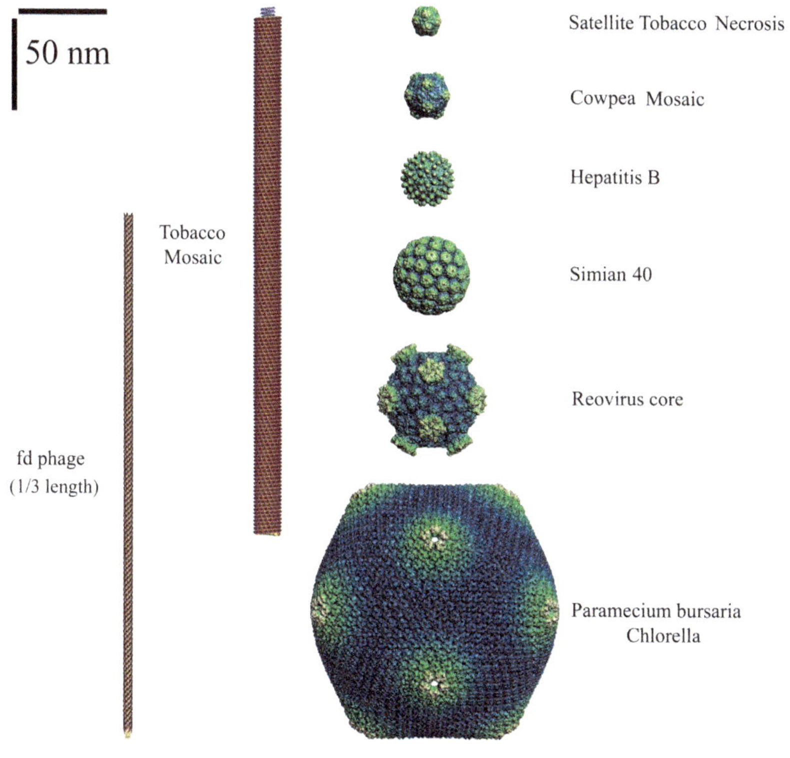

Figure 3.1 Viruses. Space-filling image reconstructions of different viral forms. Reproduced from ref. 9 with permission from VIPERdb (http://viperdb.scripps.edu).

inspiration from the virus architecture in general terms of structural consistency. Attempts have been made to exploit the geometric periodicities of viruses with a view to creating nanoscale morphologies with accuracy matching that of wild-type viruses.

There is scarce information as to functional designs, however, *i.e.* viral-like particles which would be able to encapsulate nucleic acids, deliver them into cells or tissues of interest and elicit genetic responses measurable against those of other systems. Designs probed to encapsulate genetic materials are emerging, but these often lack the demonstration of biological properties. Gene packaging is important for probing the compatibility of capsid designs with gene encapsulation: their ability to protect, transfer and release genetic cargo. Conversely, encapsidated nucleic acids, their size, type and information they carry, define the very form of resulting capsids giving rise to a fascinating diversity of viruses, Figure 3.1. From the structural point of view, the wealth of viral forms is apparent. All viruses share common features and follow similar principles in organization.[8,9] Virions can be linear or helical (tobacco mosaic virus), spherical adopting isometric symmetry (adenovirus) or of either type enveloped in a phospholipid membrane (helical influenza or icosahedral herpes simplex). Functionally, viruses exhibit the same set of common properties including infectivity, cell entry, intracellular transport and replication. Therefore, an archetypal virus can be viewed as a protein cage encapsulating a gene, capable of cell entry and endosomal escape, promoting gene delivery into the cytoplasm or the nucleus and producing offspring particles with subsequent release from the host cell. What is shown for an archetypal design has relevance for all viruses: the virus architecture offers a suitable template for artificial designs. It is symmetry driven, which guarantees the structural reproducibility of viruses and allows them to accommodate materials of different sizes and types. Wild-type viruses can accommodate different cargo without compromising their native morphology.

The principles of the structural organization of viruses are well understood and have been thoroughly described, providing a rationale for the hierarchical, bottom-up assembly of basic units into larger and orderly structures. The basics of the rationale were introduced by Crick and Watson as a theory of identical subunits.[10] The theory considers several conventions: viruses are too small to encapsulate genomes that would be large enough to encode many proteins. This is true even considering polycistronic gene expression common for viral genomes. Therefore, a virus can afford only a limited number of proteins, which in turn would be insufficiently sized to build up an entire capsid. The theory explains that multiple copies of capsid proteins are packed symmetrically and that each copy is in the identical environment of other copies. An obvious solution for such equivalent environments is to pack proteins into a cubic symmetry, which can be rotated around its axes to produce multiple even projections, resulting in a minimum free energy structure. One important consequence from this theory is that basic geometric rules can be used to describe and design virus-like objects.

Indeed, many genetically unrelated viruses adopt an icosahedral symmetry for their capsids. Human viruses (rhinoviruses, adenoviruses), plant viruses (rice yellow mottle virus, cowpea mosaic virus), animal porcine parvovirus and bacterial tailless phages belong to different taxonomic groups, but are all proteinaceous icosahedra.[8]

An icosahedron is the simplest starting template for virus-like designs. This is a Platonic solid with 20 flat faces of equilateral triangles arranged around a sphere, giving rise to a 5/2 3 symmetry, Figure 3.2a.

The entire surface of an icosahedron being applied to a protein shell would require 60 asymmetric protein copies or subunits to cover (the subunits are arranged around icosahedron vertices giving three subunits per triangle), Figure 3.2a. There are six fivefold, 10 threefold and 15 twofold symmetry axes in an icosahedron. The axes cannot be occupied by the protein subunits. Each of 60 protein subunits placed in the 20 icosahedron faces would be identical

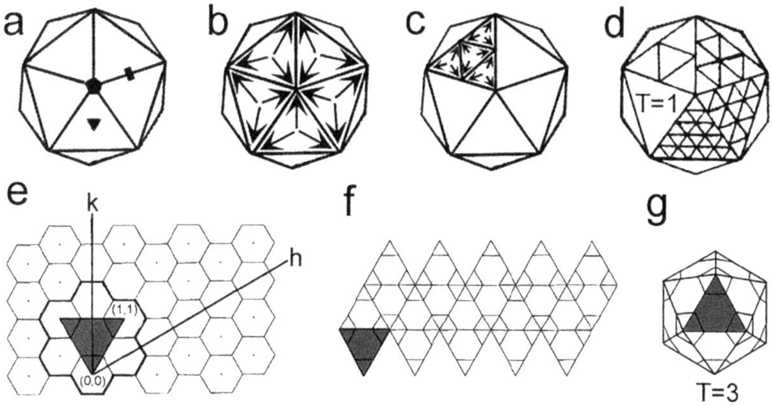

Figure 3.2 Icosahedral symmetry. (a) Schematic representation of an equilateral icosahedron, which comprises 20 flat triangular faces and 12 vertices arranged around a sphere and related by fivefold (diamond), threefold (triangle) and twofold (rectangle) axes of rotational symmetry. (b) An icosahedron accommodating 60 equivalent subunits. Proteins (arrows) are arranged around the icosahedron vertices giving three subunits per triangular face (shown only in five triangles). (c) An icosahedron accommodating more than 60 subunits. Each face is divided into four smaller facets. Proteins (small arrows) are placed into each corner of the facets (shown only in one face) giving in total 240 quasi-equivalent subunits, with 180 arranged into 30 hexamers and 60 into12 pentamers. (d) Triangulation numbers with $P = 1$ clockwise: 1, 4, 9, 16 and 25. (e) A hexagonal net lattice relating icosahedral and quasi-equivalent symmetries. An icosahedral face for a $T = 3$ surface lattice is defined in the net by the grey triangle with bold lines ($h = 1$, $k = 1$). (f) A net of 20 identical faces, with each planar hexamer composed of six triangles. Convex pentamers are generated when one of the triangles is removed. This ensures the connection of the two free edges of one pentagon in 3D with subsequent folding into a $T = 3$ icosahedron (g). Images (e)–(g) reproduced from ref. 11 with permission from Elsevier, Copyright 1997.

with its neighbours, Figure 3.2a. The subunits themselves are also symmetric and hence can be built from the smallest number of different proteins. This may explain why viruses opt for minimalistic genomes, which is in good agreement with the theory of identical subunits, and why viral genomes are polycistronic, which allows them to take the same space as would be possible for one gene translated into a single protein. Thus, viral capsids can be more complex and formed from more than 60 subunits. In fact, all viral capsids, apart from a few, tend to comprise more subunits. For instance, mosaic viruses contain 100–150 blocks and adenoviruses are thought to have as many as 252 subunits.[11,12]

However, the number of subunits that an icosahedron may accommodate is limited to 60, Figure 3.2b. This paradox required an adjustment taking into account the observations that capsid subunits are (1) non-identical and (2) symmetrical and (3) can be placed on symmetry axes.[12] The 5/3 2 rotational symmetry remains consistent with the observations provided that the subunits themselves are composed of asymmetric components. Following this assumption, Caspar and Klug (CK) introduced a concept of quasi-equivalent positions or a quasi-equivalence theory.[13] The concept postulates that (1) each protein subunit is positioned in similar but not identical environments (quasi-equivalent), (2) protein subunits are clustered in capsomers and (3) each of 20 triangular faces of an icosahedron can be subdivided into smaller triangles, facets, to accommodate more proteins, Figure 3.2c.

The number of facets is reflected by a triangulation number, T, Figure 3.2d. T can be decided graphically from an equilateral triangular net, Figure 3.2e–g, or using an equation: $T = f^2P$, where f is any integer and $P = (k^2 + kh + h^2)$, with k and h being any integer with no common factor.[11–14]

A $T = 1$ icosahedron will have 20 triangular faces, with each being assigned three identical facets. Given that one facet takes one protein, a total of 60 identical proteins will be generated, Figure 3.2b. For $T = 4$, each triangle is subdivided into four smaller triangles to place one protein subunit at each corner. This gives 12 blocks per face, which when multiplied by a total of 20 faces will give 240 blocks, Figure 3.2c. The blocks are organized as pentameric (those that are at vertices of the faces; 12 in total) and hexameric (those located at the vertices of facets; 30 in total) capsomer units. Thus, non-equivalent 60 pentamers and 180 hexamers can only be bonded in a similar or quasi-equivalent fashion. Similarly, most $T = 3$ viruses will have 180 non-identical subunits (three proteins multiplied by 60 triangles). Hence, for all $T > 1$ viruses, proteins will always have different numbers of neighbours and none in any given pair will be in a strictly equivalent environment, Figure 3.3.[9,15,16] As an exemplar, adenovirus has a $T = 25$ capsid, which equates to 1500 (60×25) subunits corresponding to 12 pentamers and 240 hexamers and incorporates at least nine different polypeptides. Thus, T numbers correlate with the number of proteins in the capsid, with larger capsids having higher T numbers.

From the design perspective, it is plausible that having lower triangulation numbers may help minimize the number of different polypeptide sequences and subunits in the final assembly. In other words, designing a $T = 1$ virus

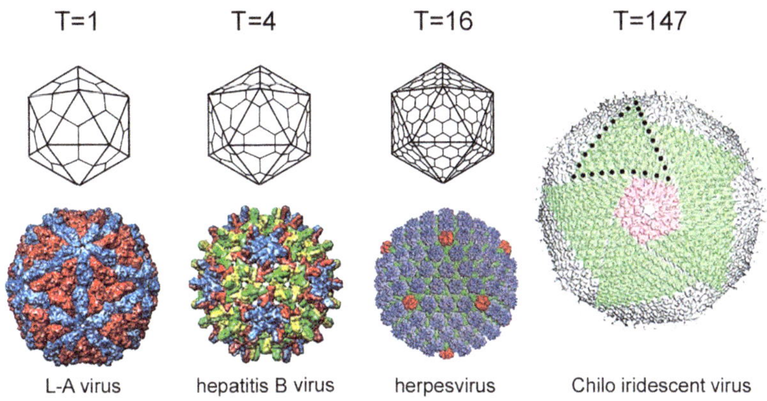

Figure 3.3 Classification of viruses according to triangulation numbers. Examples of viruses with the corresponding T numbers: $T = 1$ (L-A virus) and $T = 4$ (hepatitis B virus), reproduced from ref. 9 with permission from VIPERdb (http://viperdb.scripps.edu); $T = 16$ (herpesvirus), reproduced from ref. 16, https://doi.org/10.1371/journal.pbio.2006191, under the terms of the CC BY 4.0 license https://creativecommons.org/licenses/by/4.0/; and $T = 147$ (Chilo iridescent virus), reproduced from ref. 15 with permission from Springer Nature, Copyright 2000.

is deemed more straightforward than for $T = 100$. The most critical point in designing capsid-like structures is to ensure that all proteins within a capsomer are identical. Non-identical proteins would still adopt an icosahedral architecture as they do, for example, in polio or rhinoviruses.[17,18] However, predicting protein stoichiometry in relation to bonding structures of icosahedra is not obvious. The quasi-equivalent theory allows pseudo-equivalence relations, in which triangulation numbers of non-quasi-equivalent viruses are templated on quasi-equivalent viruses. However, this does not mitigate other complications in viral icosahedra. For instance, right- or left-handed icosahedra can still fit by this theory, but mixed or aberrant structures can not. The theory does not consider inter- and intra-capsomer interactions with different bonding strengths, and falls short of describing distorted icosahedra (*e.g.* human papillomavirus). Nevertheless, it shows good predictability in the mathematical description of virus structures and provides a firm foundation for modern taxonomy of viruses. There exist several databases built upon the theory, notable examples being ICTVdB (International Committee on Taxonomy of Viruses database)[8] and Viper (Virus Particle Explorer, Scripps Institute).[9]

The virus architecture attracts continuous interest, with contributions from different fields including those at the interface between biology and mathematics. A good number of computational models have been proposed, aiming to provide optimal descriptions for the structure of viral capsids. The concepts of internal switching and topographic potential surfaces, minimal thermodynamics and stochastic growth models, classical nucleation and tiling theories continue to advance our understanding of virus organization.[19–23] As more experimental evidence becomes available there are increases in the number of structural outliers, which cannot be described by the current models. Revisions of the existing theories have been proposed in order to offer generic design principles for icosahedral and octahedral architectures, which are described in terms of Archimedean lattices.[24] The numbers of capsid proteins and capsid layouts for viruses from different lineages and even within the same lineages prove to be incompatible with the geometric constraints of the quasi-equivalence theory.[25–27] In particular, this concerns capsids assembled from major and minor capsid proteins, the surface structures of which, however, can be formulated using a recently proposed Archimedean lattice framework, Figure 3.4.

Surface models help in devising design rules to predict the packing of capsid proteins. Complementary insights are provided by molecular dynamics simulations, which incorporate protein folding constraints into potential designs. Modern approaches combine coarse-grained and atomistic simulations to probe the stability and dynamics of designed assemblies across relevant length scales. The simulations provide better predictions to fit the experimental data, which nowadays extend to probing the behaviour of the assemblies in native and near-native environments. Thanks to the recent advances in high-resolution microscopy allowing biological systems to be visualized in water and in real time, such predictions draw upon the power of deep learning algorithms, which become indispensable tools for the prescreening of artificial designs.

Computational prediction algorithms and mathematical theories may complement one another to guide design criteria better. Most recent examples convincingly demonstrate that protein capsids can assemble from *de novo* polypeptide sequences, re-engineered natural proteins and hybrid structures. With initial attempts at engineering peptidic or glyconanoparticle aggregates, greater emphasis is now placed on obtaining finer structures: hollow, monodisperse, of controlled and uniform size, and with proven

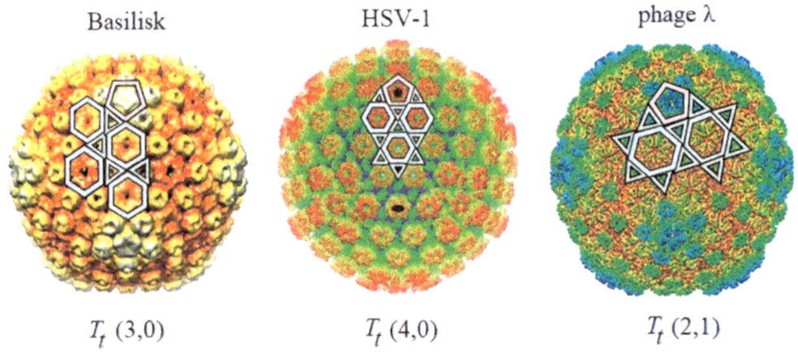

Figure 3.4 Structural outliers for viruses within the same lineage. Polyhedral surface lattices superimposed on figures showing protein units: pentagons (red), hexagons (blue) and triangles (green).[25–27] Reproduced from ref. 24 with permission from Springer Nature, Copyright 2019.

ability to deliver functional genetic material. Many designs are reported as artificial viruses,[28] although these structures are more reminiscent of particles given their spheroidal appearances, without providing experimental support for periodic surface patterns typical of paracrystalline materials, *i.e.* viruses. Most such structures are densely packed vesicles or micelles rather than delineated deltahedra. With new measurement techniques able to distinguish hollow from solid morphologies, assembly rules are being revised to predict the assembly of hollow, enclosed shells with a defined cavity,[29] with increasing characterization requirements for viral mimetics.[30] Other viral geometries are also exploited for nanoscale designs: protein shells templated on truncated icosahedra and polyhedra prove to exhibit virus-like structural and biological properties.[31] Further inspiration is taken from filamentous virions, including filoviruses, plant mosaic viruses and bacteriophages.[32–34] Filamentous viruses are linear and tubular morphologies adopting helical periodicities in their assembly. Tobacco mosaic virus (TMV), first shown to self-assemble *in vitro* from purified components,[35] has served as a comparison material for structural studies of viruses, Figure 3.5a.

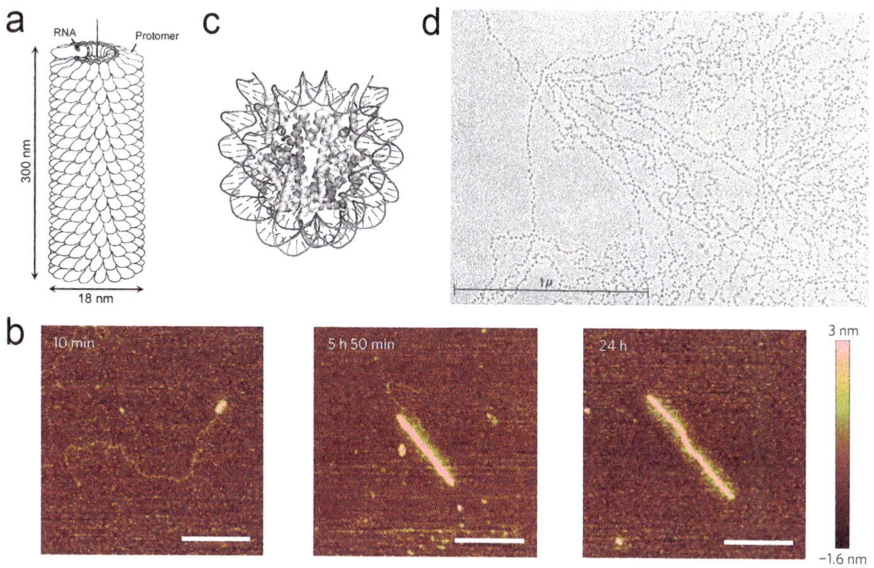

Figure 3.5 Filamentous virus and nucleosome assemblies. (a) Schematic representation of a TMV rod assembled from protomers arranged around RNA. Adapted from ref. 40 with permission from Elsevier, Copyright 1963. (b) Atomic force micrographs of filamentous virus-like particles assembled from a designed protein wrapped around a linear double-stranded DNA. Different incubation times are indicated in the images. Scale bars are 200 nm. Reproduced from ref. 45 with permission from Springer Nature, Copyright 2014. (c) Structure of the nucleosome core particle, down view with respect to the DNA super-helix axis (PDB 1P3A). (d) Electron micrograph of nucleosomal beads-on-a-string arrays. Adapted from ref. 48 with permission from Elsevier, Copyright 1976.

TMV is composed of about 2130 identical protein subunits, called protomers. By winding around a single-stranded RNA, protomers stack with the formation of a rigid rod.[36] The assembly initiated by a specific lollipop-shaped region in the viral RNA interacts with protomers to form a disc as an assembly unit. Discs stack together to elongate into a rod. In the process, RNA with the protein subunits dislocates into a proto-helix which closes the axial gap between the discs and buries RNA inside the rod. This arrangement proves optimal as nucleic acids are prone to adopt helical conformations, ensuring identical bonding patterns between the RNA and the subunits, and supporting the cooperative assembly of discs.[37] This is a good example of how the physical dimensions of a virus depend on those of the encapsidated nucleic acid. The helix is a conserved structure. The diameter and width per turn are 8 and 2.3 nm, respectively. The turn propagates in the fully assembled virus to a 280–300 × 18 nm tube. The tube has a central pore or channel with a uniform diameter of 4 nm that is supported by the RNA, which as a scaffold associates with the protomers near their inner surfaces. The RNA helix lies in the interstices of adjacent turns of the helix formed by the protein protomers, thus shaping a single linear structure of two mutually supporting helices.[38–40] The two ends of the tube are not equal: one end supports end-to-end elongation and is said to be active, whereas the other end does not propagate and remains inactive, which limits the formation of higher order structures to dimers.

M13 bacteriophage is another filamentous virus popular in nanoscale designs. It is a single-stranded DNA virus with its capsid assembled from three different types of proteins: 2700 copies of a major coat protein are arranged in a repeating helical array around the DNA to shape a flexuous cylinder. The cylinder is longer (860 nm) and thinner (6.3 nm) than the TMV tube and is capped at both ends by the other two proteins. The size of the DNA appears to regulate the copy number of the coat protein. In a microphage variant of M13, an assembled cylinder is shortened to 50 nm. Only 95 copies of the protein are needed to coat the DNA, which is 30 times shorter than the size of the wild-type strand. This property is common for filamentous viruses and can be empirically emulated for designing size-controlled virus-derived nanostructures.

Others filamentous viruses that exhibit similar exploitable characteristics include M13-related phages, such as Enterobacteria fd phage, Figure 3.1, plant viruses (Flexiviridae) and pleomorphic primate filoviruses.[41,42] Many have other unique and useful properties and some are being considered for different nanotechnology applications. In comparison with spherical viruses, filamentous viruses also represent more straightforward models to understand better the principles of gene packaging. The advent of more effective gene delivery and gene editing technologies and systems, such as CRISPR, created new possibilities for targeting complex genetic diseases that require the use of large DNA. Similarly, the rapid progress in synthetic biology, which looks into how living cells can be repurposed into manufacturing platforms, prompted the adoption of mammalian cells.

These cells hold promise as more cost-effective platforms by offering much higher metabolic capacities than bacteria and yeasts. However, to realize their potential fully, intracellular delivery of large DNA up to the chromosomal size is required. Gene packaging in natural viruses relies on the highly cooperative assembly of protein subunits on nucleic acid templates. Protein binding at a specific origin-of-assembly region on the nucleic acid template initiates the assembly. Upon binding, the protein undergoes a conformational switch, enabling it to complex with its genome. The complexation must be reversible to allow the release of the viral genetic material into the cytoplasm or nucleus of the cell. Often overlooked in synthetic designs, this property is important for genetic reactions, be it gene silencing or transgene expression. For example, it has been shown that polypeptides with cell-penetrating properties bind to RNA in a folding-responsive manner, and may conformationally lock it, thus preventing its release and functioning.[43] Perhaps for these reasons, natural viruses also use allosteric protein–gene interactions and allocate scaffolding proteins to support the assembly, in particular that of large viruses.[44] Along these lines, a kinetic model for filamentous viruses has been proposed. Inspired by the TMV coat protein, a minimalistic artificial protein was engineered to complex with a linear double-stranded DNA, resulting in rod-shaped virus-like particles.[45] The protein comprised three specialist domains: one to bind electrostatically to the DNA, a second to prevent lateral aggregation of forming complexes and a third, central, domain to provide control over the cooperativity of the assembly. This third domain was designed as a sequence of short peptide repeats, which helped in determining empirically the threshold of cooperativity in the formed complexes. Remarkably, at defined charge ratios of DNA to protein, the virions could assemble into monodisperse and uniform particles, with each particle containing just one DNA molecule, Figure 3.5b.[45]

This and similar designs offer promising strategies for self-instructive virus-like systems. Most strategies to date, however, have been demonstrated for relatively small genes, which can include plasmid DNA, but fall short of cooperative complexation with larger nucleic acids. To expand the design space of artificial systems, fundamentally different or somewhat complementary strategies are needed. Unlike prokaryotes, the preference of which for small genomes is driven by metabolic and spatial economy and also the speed of cell multiplication, eukaryotic cells preserve their secondary DNA. This carries the cost of devising an ingenious solution for gene packaging. Indeed, eukaryotic DNA is much larger than viral genomes and is not space compromising. The human genome contains approximately 3 billion base pairs packaged into 23 chromosomes. Since most cells are diploid, *i.e.* have 23 pairs of chromosomes, each cell contains 6 billion DNA base pairs. With each base pair spanning 0.34 nm, there is over 2 metres of DNA in each cell and 100 trillion metres of DNA in the human body, which contains 50 trillion cells. Remarkably, however, 2 metres of DNA is packed in the cell nucleus with a size of just several cubic micrometres.

To mitigate such a discrepancy between the DNA size and the capacity of prokaryotic gene-packaging systems, eukaryotic cells use nucleosomes.[46–48] These structures rely on similar principles of symmetry-driven assembly of protein building blocks. However, instead of building a complex protein coat, the encoding of which would require a considerable space of the genome, eukaryotic DNA winds on multiple copies of symmetrically arranged protein spools called histones. DNA is wrapped around by these bobbins, eventually becoming caged in spiral DNA nanotubes, Figure 3.5c.

Histones embrace themselves with multiple rings along the double-stranded DNA that are held from inside and are stacked in a spiral fashion. This mode of assembly is structurally the reverse of that of viruses. Here it is proteins that become encapsulated by nucleic acids. Taken liberally, nucleosomes can be thought of as inside-out viruses. Like viruses, they are regularly built capsules and their encapsulated components guide the assembly and morphology of the capsule. Unlike viruses, nucleosomes are not discrete assemblies. They constitute an initial phase of assembling into more complex structures – chromatins.[49] Briefly, 11 nm thick DNA nanotubes folded by histones coil further to form 30 nm wide filaments. Involved in cellular processes, these filaments fold reversibly without causing structural tensions to mechanically rigid DNA. The rings of nucleosomes are flexibly linked, ensuring rotating mobility. Decompacted DNA fibres are intermittently knotted threads, termed "beads-on-a-string", Figure 3.5d.[48] The chromatin assembly supports functional well-choreographed processes. For example, genes that are coiled or spooled into nucleosomes are locked or switched off, and can be activated as nucleosomes are site-specifically reassembled.[50] Nucleosomes can be induced to move along the DNA string and their distribution on the string can be controlled.[51] However, there is a great deal of detail yet to be revealed in the nucleosome assembly before meaningful nucleosome-inspired designs can be realized. Interestingly, probing nucleosomes for gene integration is starting to gain some traction. Recent examples include retroviral integration into nucleosomes and the design of human artificial chromosomes.[52,53]

3.2.1 Triskelion Topology as a Template for Cage-like Assemblies

Viruses opt for encapsulation to deliver their genetic material safely to host cells. The intracellular delivery of viruses, as of any other material type, also occurs *via* an encapsulation process – endocytosis.[54] There are several types of this process that have been extensively studied and are very well covered in the literature. The process itself relies on the flawless performance of several encapsulating systems, which follow the same principles of symmetry-driven assembly. Clathrin is one such system.[55] Clathrin is recognized for its unique shape, a triskelion or cross-like structure with three bent legs, Figure 3.6.[56] Adaptor proteins recruit the triskelion onto membranes, promote the assembly of clathrin into polyhedral lattices or cages and promote the inclusion of membrane-bound cargo in membrane vesicles. The lattices then coat a new vesicle that encapsulates the cargo for transportation.[57]

As with the virus architecture, the assembled clathrin coats exhibit regular arrangements of open pentagonal and hexagonal facets, Figure 3.6. The exact structure of the coats is specified by the process they support: an outside coat is typically 70–200 nm in diameter, with triskelion hubs located at the vertices of its lattice, which are separated at 16 nm. Each triskelion leg has a bending site that allows it to extend around two edges of pentagonal and hexagonal facets protruding inwards, Figure 3.6. Each bending site is introduced by a knee and a linker. The knee creates a strong twist, dividing each leg into proximal and distal segments, which fit together with legs of other triskelions to form the edges of the facets. Different combinations of interactions between legs are possible, *i.e.* distal–distal, proximal–proximal and distal–proximal. Generally, three proximal segments of a triskelion extend towards three adjacent vertices, with distal segments running to the next vertices. For instance, a proximal segment of one triskelion extends along that of an adjacent triskelion. At its knee, the leg of one triskelion curves so that its distal segment lies just beneath the proximal segment of a second leg of the adjacent triskelion, Figure 3.6. The knee of a triskelion is placed at an adjacent vertex, where it primarily serves to help the triskelion to accommodate the formation of pentagonal and hexagonal facets arranging coats of various diameters and curvatures.[56] Differences in leg conformations in individual clathrin barrels are mainly attributed to the flexion regions near the

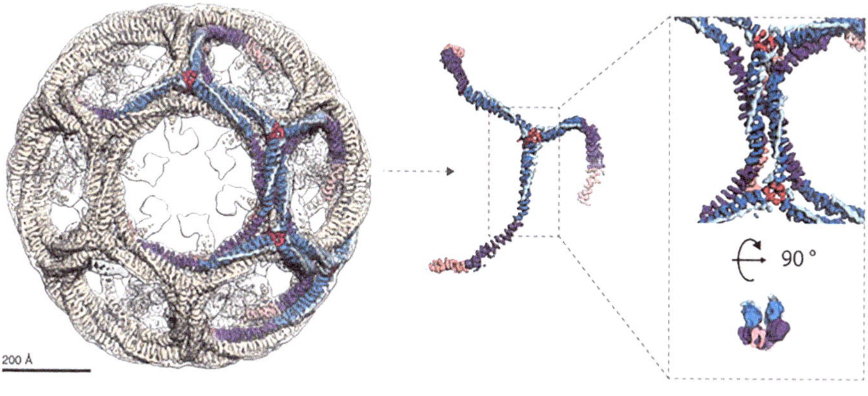

Figure 3.6 Clathrin. Cryo-electron microscopy map of a mini-coat clathrin architecture (left) with four triskelia highlighted to show quaternary-level interactions. A triskelion is shown in isolation (middle) and with neighbouring heavy chains (right). Structural features are coloured according to the domain structure of the clathrin heavy and light chains. Reproduced from ref. 56 with permission from Springer Nature, Copyright 2019.

vertex. The flexions may be responsible for additional capacities inherent to clathrin cages, *e.g.* the legs can relax local strains associated with entrapping asymmetric or large cargo without the reassembly of coats.[58]

Adaptor proteins that promote the assembly of triskelions into lattices create an interface between an outside coat and an enclosed membrane by connecting the membrane with N-terminal domains of each leg. A flexible segment of the leg precedes the terminal domain, which together with a linker between the distal and terminal domains gives the leg a smooth continuous turn. The linker also helps position a foot-like terminal domain roughly perpendicular to the main axis of the leg, Figure 3.6. Collectively, these structural features arrange a flexible, binding-responsive three-dimensional triskelion with a height of 20 nm.

Similarly to viral capsids, the assembly of clathrin coats is symmetry driven and uses a basic dihedral topology assuming sixfold space group symmetry, *i.e.* a hexagonal lattice. Twelve pentagon facets are required to close the lattice, with a dominant proportion of hexagonal barrels assembled from asymmetric units. Each unit is composed of three triskelions or nine legs, which are aligned in the proximal segments and diverge at the knees.[56,59] Different degrees of bending in flexion regions support altered positions of distal segments and terminal domains relative to the vertex, but without compromising on structural compliance and hence similarity of all segments along the leg. Triskelion legs extend as curved 48 nm rods approximately 2 nm in diameter, which is consistent with the proximal, distal and linker segments appearing as zigzags of α-helices.[55,56] Five zigzags (140–150 residues) constitute a repeat pattern with no interruptions in the leg. Adjacent zigzags cluster reversibly into four-helix bundles, which allows them to maintain larger gaps between successive zigzags, the accumulation of which endows the leg with a smooth curvature.

Similarly, the structure of the terminal domain is well adapted to support its main function by forming an extensive surface of protein interactions. This domain folds into a seven-blade β-propeller. The blades are arranged around a central pseudo-tunnel closed by tight contacts between terminal blades. Many proteins use this type of folds to mediate complex interactions in constrained environments: as interior elements of protein architectures, β-propellers provide structural support for specific binding and recognition patterns. Often found in enzymes, the structures create suitable environments for biochemical transformations or serve as scaffolding platforms for regulatory multiprotein complexes.

This structural perfection of clathrin architecture compounded by its functional efficacy opens up exciting possibilities for engineering nanoscale systems. Clathrin assembly is very effective and rapid and is completed within 1 minute from membrane binding and cargo entrapment to coat formation and vesicle transport with the subsequent removal of clathrin coats. However, a functional clathrin mimetic has yet to emerge. The structure of the protein is not trivial, and its sequence is highly conserved. Imitations of triskelion structures are more common, including small-molecule triskelions, which are more synthetically accessible, and triskel conjugates of peptide and nucleic acid blocks.[60–64]

The clathrin triskelion represents an old form that remains challenging to replicate fully by design, but it also offers a relatively simple topology inspiring creative solutions for virus-like designs. First, triskelions are discrete, covalent subunits with predefined directionality of assembly. Second, each triskelion arm provides a complementary binding surface for the arm of another triskelion. This type of binding inherently supports rapidly propagating patterns, which are diverse in size and shape, as observed in clathrin coats. In contrast, the symmetry of viral capsids requires protein subunits to arrange around five- and sixfold vertices, suggesting that capsid proteins must interface *via* precisely matched orthogonal interactions, which in itself is a major design challenge.

The triskelion topology can help address the problem of orthogonal fitting by placing the same protein motif at the triangular vertices of the capsid and retaining the threefold symmetry of triangular faces. A rigid C_3-triskelion occupying a triangle provides a practical application for this strategy.[65] Two triskelion arrangements within a triangle are possible, leading to two different propagation modes, "starburst" and clathrin-like "honeycomb", Figure 3.7a. Each of these arrangements conforms to a classical 5/2 3 capsid

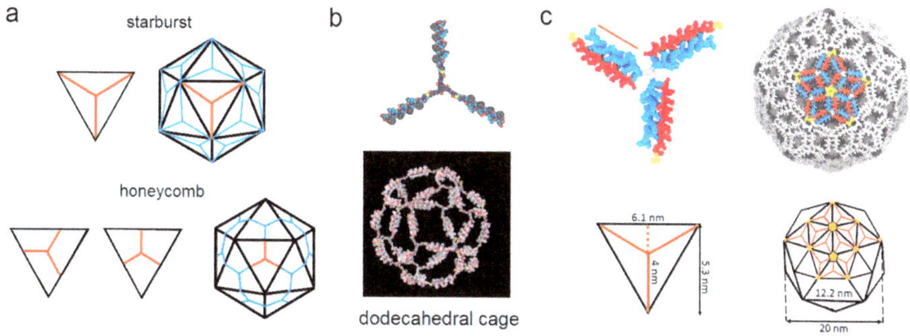

Figure 3.7 C_3-triskel assembly topologies. (a) Two arrangements of a C_3-triskelion within a triangular unit leading to "starburst" and "honeycomb" propagation modes. In the starburst mode, the three equivalent legs of the triskelion serve as the altitudes of the triangle. Two variations of the honeycomb triskelion arrangement, symmetrical and asymmetrical, are given. For clarity, central triskelions are shown in orange. Reproduced from ref. 65, https://doi.org/10.1038/s41467-017-02475-3, under the terms of the CC BY 4.0 license https://creativecommons.org/licenses/by/4.0/. (b) The "honeycomb"-like assembly of peptide C_3-triskelions *via* β-sheet formation into a virus-like sphere. Reproduced from ref. 67 with permission from Springer Nature, Copyright 2012. (c) An α-helical "starburst"-like C_3-triskelial assembly. The molecular model is based on PDB entry 4DMD, rendered by MacroModel Schrödinger. Each subunit of the triskelion corresponds to the altitude of an equilateral triangle. Each arm of the triskelion spans 4 nm, which makes the side of the hosting triangle 6.1 nm. Reproduced from ref. 65, https://doi.org/10.1038/s41467-017-02475-3, under the terms of the CC BY 4.0 license https://creativecommons.org/licenses/by/4.0/.

symmetry, and each requires the triskelion to propagate at its termini. The "honeycomb" propagation relies on interfacial complementary interactions, which can be oblique or orthogonal.

Exemplar designs used re-engineered naturally occurring peptide motifs, adopting β-pleated structures. The clathrin-like mode of assembly in these designs is preferred owing to the antiparallel arrangement of β-strands. For example, a sequence design mimicking virus-derived tryptophan zippers, with a characteristic core motif WTW, adopts β-form conformations with cross-strand tryptophan rings packed tightly against one another.[66] A triskel structure of this sequence folds as a β-turn and pairs with another arm of another triskelion with the formation of a β-sheet.[67] Triskelions then propagate into two-dimensional sheets with increasing curvature, which facilitates the formation of three-dimensional hollow shells, Figure 3.7b. Like clathrin lattices, these shells can be polydisperse in size but are uniform in shape. Other β-structured triskelion designs furnish similar cage-like assemblies, and all of these designs exhibit the level of structural plasticity characteristic of the triskelion topology. Sequences adopting helical conformations have also been shown to assemble *via* the clathrin-like "honeycomb" mode.[61,68,69] However, these sequences were not constructed into covalent triskelions. Instead, they formed three-arm branching units which propagated at twofold axes of symmetry. The formation of disulfide bridges in these designs proved to be important as an auxiliary means for structural stabilization. For the clathrin-like mode of assembly, strong binding at twofold axes is crucial because five- and sixfold vertices remain hollow and redundant. Therefore, this mode of propagation is energetically favoured for short-legged triskelions or orthogonal C_3 branching units which can afford tight packing. The "starburst" arrangement is better suited for capsid-like assembly in that it is enabled by all three C_3 termini clustering at five- and sixfold axes, and therefore can readily translate into any $T > 1$ virus-like assembly, Figure 3.7c.[65] The three equivalent legs of each triskelion provide the three altitudes of a triangle. Therefore, the edge of a triangular unit is a common integer for all units in an icosahedron face, irrespective of their number. Using these principles, a *de novo* triskelion was designed to assemble into virus-like capsids, Figure 3.7c. Since each triskelion must adopt a rigid conformation to allow it to fit into a triangular face of the capsid template, a complementary, counterpart sequence was designed to dimerize with each of the triskelion legs. The sequence has a terminal cysteine residue incorporated to activate the clustering of the folded triskelions at their leg termini, stabilizing the assembly in solution into predominantly $T = 4$ capsids, 20 nm in diameter.[65] This design also is a notable example of providing new functions that are not typical of either clathrin cages or viruses. The assembled capsids attack bacterial membranes in which they rapidly convert into expanding pores, causing membrane lysis in minutes. Hence the design demonstrates the first instance of antibacterial capsids that execute a new primary function by destroying bacteria on contact.[65]

3.2.2 Cage-like Encapsulation for Nanomaterial Synthesis

Molecular encapsulation for transport is a unifying purpose for clathrin and viruses. However, these structures and protein shells in general are not limited to this function only. Other applications are extensively being tried so as to benefit from the precision of nanoscale encapsulation. Proteins as the main building blocks of native systems remain the preferred candidates for encapsulating host systems, which is justified by the versatility and abundance of protein shells. The types of guest molecules prove to be more diverse, ranging from inorganic materials to macromolecules other than nucleic acids. Metal ions and minerals are ubiquitous guests. A classic example of protein shells specialized in mineral encapsulation is the protein–iron complex ferritin.[70]

Ferritin is a conserved protein assembly assigned to store and transport iron to and from the cell cytoplasm. The protein part of the complex is referred to as apoferritin, which assembles into a closed 12 nm sphere with a nearly spherical cavity 8 nm in diameter, Figure 3.8a and b.[71] Iron is fixed in the formed cage in the form of microcrystalline ferric oxyhydroxide phosphate, $[FeO(OH)]_8[FeO(H_2PO_4)]$.[72] The mineral contains thousands of iron atoms protected from complexation with water. Apoferritin thus preconcentrates iron to the levels required by living cells. The apoferritin shell comprises 24 polypeptide subunits, each folding as a four-helix bundle. There are two types of the bundles, heavy (H) and light (L). The ratios of these two types in ferritin shells determine structural variations in ferritins from different organisms and tissues and differences in the rates of microcrystalline formation.[71–74] Ferritins are dodecahedral shells, in which 12 rhombic faces are shaped by 12 dimers of polypeptide subunits, Figure 3.8b. The dodecahedron structure confers all of the properties of ferritins. For example, subunits in each dimer form grooves which cluster into a network lining the inner side of apoferritin. This network is thought to nucleate the formation of an iron core, while channels formed at the intersections of three and four subunits render the assembled cage holey or porous, Figure 3.8b. Ferritin controls in-and-out transport of iron through these channels, which can be three- or fourfold and are similar in size (0.3–0.4 nm), but have different polarities.[71] Anionic amino acids (aspartates and glutamates) and hydrophobic residues (leucines) occupy three- and four-fold channels, respectively, to support two different functions of ferritin.

The sequestered iron in ferritin is concentrated as a ferrihydrite. This is a crystalline lattice, in which each Fe(III) species is caged by five oxygen ions and a phosphate.[74] Phosphate groups bind residues of the inside of the apoferritin shell and hence are located on the outside of the microcrystalline core. Iron atoms cannot be released from the lattice unless it is broken. To dissolve the lattice, ferritin reduces iron from the ferric form Fe(III) to the ferrous form Fe(II), which is then entrapped by water molecules to become hydrated and soluble, $Fe(H_2O)_6^{2+}$. In this form, it can be released into

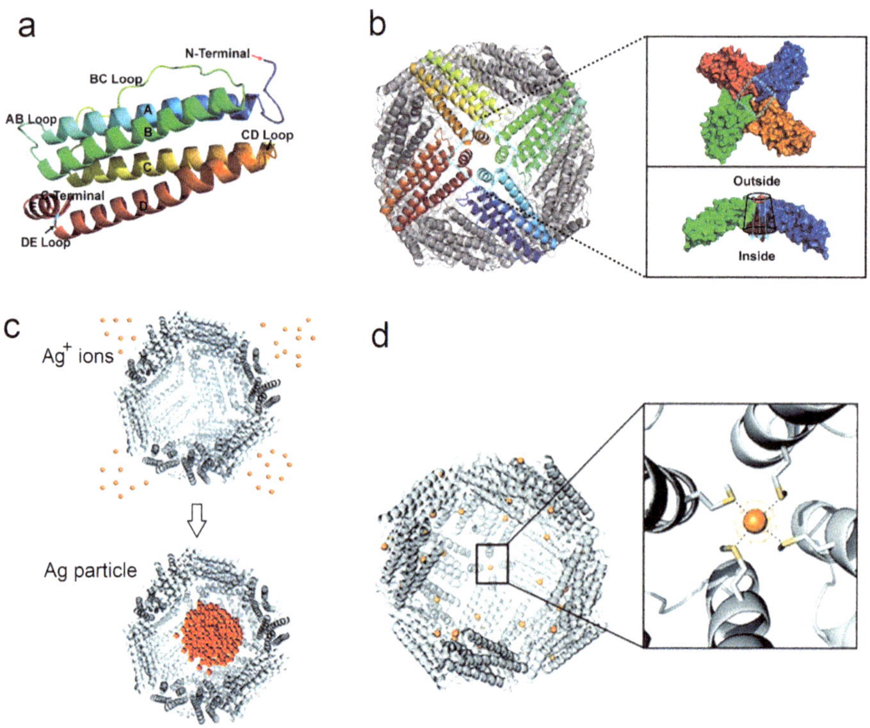

Figure 3.8 Ferritin as a template for nanomaterials. (a) The building block of the ferritin subunit and (b) a ferritin cage facing up with a fourfold channel. Reproduced from ref. 71 with permission from Elsevier, Copyright 2020. (c) Formation of silver particles from silver ions *via* a metal redox reaction in an 8 nm apoferritin cage. (d) X-ray crystal structure of an Ag(I)-containing ferritin. View along the fourfold channel. Four chains were removed to display the internal cavity of the protein, with a silver binding site shown as an enlarged image. S and C atoms are shown in yellow and grey, respectively, and silver molecules are shown as orange spheres. The silver atom is coordinated by Met153 from four symmetry-related subunits. Adapted from ref. 73 with permission from American Chemical Society, Copyright 2010.

the cell cytoplasm.[72,74] Cationic hydrated iron species favour interactions with anionic threefold channels, supporting the passage through the channels. Hydrophobic fourfold channels are not accessible to ferrous ions, but are believed to mediate the transport of electrons from outside the protein shell into the core. This serves to help reduce Fe(III) or solubilize the crystalline lattice for ferrous ions, which are then released by threefold channels.

Ferritin is the most studied natural system that reversibly converts metal ions to minerals at the nanoscale. Given the relative ease with which ion–solid transitions can be performed in apoferritin, it continues to be a model of choice to explore other metal conversions.

The ultrastructure of self-assembling encapsulants, in contrast to covalent systems, is able to adjust in response to changes in intramolecular dynamics and allosteric effects. Adaptive conformational switching impacting on nanoscale environments is necessary for chemical conversions, which often require accurate stoichiometry of reacting materials, but which are also prone to errors in rigid and non-rectifying ultrastructural environments. Successful encapsulants are able to adapt appropriate geometries to minimize potential errors. As extreme cases, assembly mutants or misassembled encapsulants lose assigned functions completely and can lead to further complications of aggregation. This strong dependence of self-assembling encapsulants on reproducible assembly modes prompts complementary control over the size and structure of guest components. Therefore, naturally occurring encapsulants are being exploited as nanophase reaction vessels, which has already given rise to a wide variety of bioreactors, capsules and containers.[75]

3.2.3 Native Protein Shell and Virus Templates

The ease of use, predictability and *in vitro* reconstitution of nanoscale encapsulants are the factors considered to define the preference of one type over another for templated synthesis. Although these qualities are common for natural systems, most understood protein cages and viruses provide the best represented templates.

For example, the use of ferritins for nanomaterial syntheses started from the observation that apoferritin can encapsulate metal ions other than iron. Among many, Ni^{2+}, Cd^{2+}, Mn^{2+}, UO^{3+}, Tb^{3+} and VO^{4+} bind specifically in the core sites of the apoferritin cage, Tb^{2+} and Zn^{2+} compete with iron, and Fe^{2+} was shown to displace Mn^{2+} and VO^{4+}.[75–77] *In situ* saturation and subsequent oxidation of Fe^{2+} to Fe^{3+} was found to generate new binding sites for Mn^{2+} and VO^{4+}, indicating that the core is formed in the interior of the protein.[78] These and other studies provided incentives to use different ions at various ratios to seed and control the formation of native and foreign cores, Figure 3.8c and d.[73,74]

First attempts were reported in the early 1990s. As a an exemplar, *in vitro* reconstituted ferritins were used as biological containers, able to generate mineral cores with sizes and crystalline patterns identical with those of native iron cores.[79–81] Empty and iron-loaded apoferritins could be used: *in situ* reactions of native ferrous cores with sulfides (H_2S/Na_2S) gave crystalline particles of iron sulfide, and *in vitro* reconstitution of Fe(II) with ferritin in the presence of phosphate, arsenate, molybdate and vanadate could produce similarly sized particles.[74,82] Iron oxide magnetites provided "magnetoferritin" – a semisynthetic protein hosting ferromagnetic particles.[83] In all such processes, iron uptake is a limiting stage. Iron entry can be arrested by blocking apoferritin channels, *e.g.* with a competitive inhibitor, suggesting that site-specific modifications of the apoferritin interior can endow these old and well-exploited natural forms with new functions.[74] For example, the channels of apoferritin cages can display silver-binding sites supporting the formation of colloidal silver, Figure 3.8b and c.[73,84] Similarly, short peptides found to recognize specific metal ions selectively could be used to decorate the interior of apoferritin cages, providing new prescription cages for the synthesis of desired nanomaterials.[85–87] Attempts to use apoferritin cages themselves as building blocks for larger materials have also been reported. Outer surfaces of the cages coated with biotinylated ligands self-arrange into three-dimensional networks by interaction with a biotin-binding protein, streptavidin. Such an approach yields ordered arrays of ferritins amplifying the effect of a single protein, which may find use in microelectronics or nanostructured batteries as energy storage units.[88,89] Enhanced power density, superior storage and discharge rates and compactness are among the desired properties that are new to both the protein and the application in which it is used. Building upon these attempts may provide more cost-effective technologies and environmentally friendly energy sources. Ferritins can be repurposed as endogenous reporters or contrast agents for non-invasive magnetic resonance imaging (MRI).[74,90] An example is to use ferritins to map gene expression in tumour cells in response to antibiotic treatments regulating the expression of modified ferritins.[91] Tumour cells could be identified by MRI upon an increase in iron uptake and hence in ferritin expression, whereas antibiotic treatments cause decreased levels of expression. These results suggest a means for controlling the expression of therapeutic genes monitored by MRI.[90,91]

Ferritins remain the system of choice for inorganic material conversions. Viruses offer a richer repertoire of encapsulating templates, which can mitigate the limiting setting of apoferritin cages in shape and size.[9] Viruses find use in the fabrication of semiconductors and liquid crystals, nanowires and nanotubes, and mineral films and fibres.[92] Templating nanomaterials is driven by the morphological type of the viruses: icosahedral forms are favoured for particulate or array materials in a manner similar to that of ferritins, whereas helical and rod-shaped viruses are preferred for anisotropic structures. The same structural considerations as used for ferritin-based templates are used for viruses. In addition, owing to safety concerns, a common preference is for non-infectious viruses such as TMV.

A great deal of effort has been devoted to the use of two plant viruses, namely cowpea chlorotic mottle virus (CCMV) and cowpea mosaic virus (CPMV),[93–95] Figure 3.9a. These viruses were the first icosahedral viruses to be reconstituted *in vitro*, with both being remarkably stable, exhibiting reproducible modes of assembly that are responsive to external stimuli. CCMV can increase in size by 10% by swelling *via* the formation of 60 separate 2 nm gates on the virus surface. This property to swell allows CCMV to control the encapsulation of three single-stranded RNA molecules, which are encapsidated separately. Similarly to apoferritin, the shell of CCMV appears perforated, allowing for molecular gating, Figure 3.9b. The gating of CCMV is

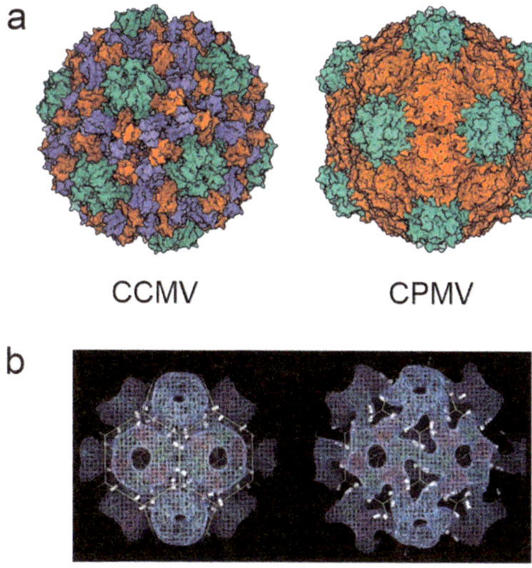

Figure 3.9 CCMV and CPMV. (a) CCMV (left, 1ZA7) and CPMV (right, PDB entry 1NY7) capsids showing subunit organization. Different colours are used to show the domain organization of pentamers and hexamers. (b) Cryo-electron density maps of the CCMV capsid at pH 4.5 (left) and pH 7.5 (right, swollen form). Reproduced from ref. 93 with permission from Elsevier, Copyright 1995.

reversible and responsive to changes in pH, Figure 3.9b.[95] Thus, this swelling mechanism, characteristic of the old form, can be exploited *in vitro* for new functions, particularly in combination with its ability to reassemble fully without the RNA components.

CCMV has a $T = 3$ capsid built from 180 identical protein subunits that arrange into a 29 nm wide particle, Figure 3.9a. The interior radius of the particle averages 10 nm, but cationic portions of the coat proteins project to 9 nm, which creates a basic cavity 18 nm in diameter.[93,95] This is approximately twice the inner diameter of the apoferritin cage, suggesting the applicability of the gating mechanism of CCMV for mineral growth. Indeed, the first attempts used CCMV as a dynamically transient host–guest system for pH-dependent inorganic polymerization.[96] A series of reactions were performed on emptied virions, which were incubated with aqueous tungstate (WO_4^{2+}) at pH > 6.5. Under these conditions, the virus is in its open or swollen form. At lower pH (pH < 6.5), tungstate ions were converted into water-insoluble paratungstate polyanions ($H_2W_{12}O_{42}^{10-}$), concomitantly causing the closure of the pores. The basic cavity of virions facilitated the entrapment and subsequent crystallization of the polyanions. Single-crystal materials were formed within the confines of viral cages. The role of electrostatic host–guest interactions for such conversions was addressed in other studies. Protein segments that shape the interior of the virus were mutated by replacing basic lysine and arginine residues with glutamates. This rendered the core acidic, making it reminiscent of that of native apoferritin. Not only had the mutants assembled into wild-type virions, they were also shown to mimic ferritins in hosting the formation of crystalline ferrihydrate nanocrystals.[97] Cysteine mutants of exterior portions of the same proteins promoted the self-assembly of CCMV into monolayers on amorphous gold surfaces. Interestingly, only capsids decorated on one side and chemically passivated on the other side were able to self-assemble into extended arrays, whereas virions with symmetrically distributed cysteine residues tended to aggregate.[98] These early examples demonstrate the use of site-specific mutagenesis, which together with chemical modifications of viral surfaces provided an efficient tool for creating addressable nanoscale building blocks.[99] The properties of the other cowpea virus, CPMV, proved to be particularly instrumental in this regard.

CPMV is an icosahedral $T = 3$ cage assembled from 60 asymmetric units each containing two subunits, large and small, Figure 3.9a. A pore centred at each fivefold axis of the cage is formed by five units. Although the exact role of the pore remains unknown, its tiny size is attributed to preventing the diffusion of materials larger than 1 nm as no reaction takes place between marginally larger (1.4 nm) gold nanoparticles and the surfaces of wild-type capsids, stripped off cysteine residues.[100–102] In contrast, native viral particles in reaction with ethyl mercury phosphate readily formed adducts that were under 1 nm in size. X-ray diffraction analysis of the loaded virions revealed that the reagent reacted with single cysteine residues located in the interior surface of the large viral subunits. Furthermore, an organic dye, fluorescein, was found to condense with the same residue in a concentration-dependent manner, with 60 attached dye molecules detected in the virus, which indicates that all interior cysteine residues are equally reactive. Capitalizing on these findings, exterior mutants of a single cysteine insertion were produced and treated with increasing concentrations of rhodamine, a red fluorescent dye. The interior cysteine remained active in mutants, which prompted the sequential derivatization of inner and outer surfaces with different molecules under controlled conditions. Mutant particles decorated with both dyes were produced.[103] The exterior of CPMV proved to be accessible to surface decorations with various molecular types constituting targeting or homing groups, therapeutic, shielding and imaging groups to provide a repertoire of specialist nanoparticles for cancer diagnostics and therapies.[104]

The observation that viral capsids can assemble independently of their genomes inspired the use of inorganic nanoparticles to template the assembly of virus capsids – an approach the reverse of those discussed thus far.[105] As an exemplar, functionalized gold particles imitating the interactions of native RNA can be used to initiate the assembly of virus capsids, which adopt the native icosahedral packing exhibiting pH-induced swelling transition. By varying the diameter of the gold particles, it proved possible also to control the structure of the assembled capsids and their shape. Viral coat proteins can be used to assemble around gold rods as opposed to gold particles. The proteins effectively assemble with the formation of spherocylindrical shells encapsulating the rods, Figure 3.10a.[106]

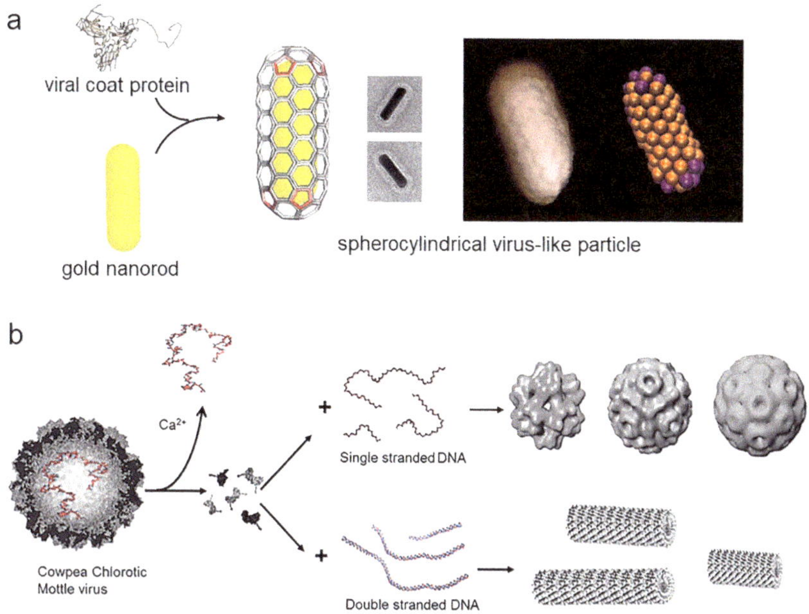

Figure 3.10 Templated reassembly of capsid coat proteins. (a) The coat proteins of brome mosaic virus assemble on gold nanorods with the formation of spherocylindrical virus-like particles, in which cylindrical sides follow hexagonal close packing and spherical caps contain pentamers. From left to right: schematic representation of the packing, electron micrographs showing gold nanorod core encapsulated in the protein coat, and atomic force micrograph showing the assembled virus-like particle and its simulation. Adapted from ref. 106 with permission from American Chemical Society, Copyright 2018. (b) Schematic representation of the reassembly of CCMV coat proteins on different nucleic acids, single- and double-stranded DNA, resulting in spherical and rod-like virus-like particles. Reproduced from ref. 107 with permission from Elsevier, Copyright 2019.

The cylindrical sides of the shell were found to be hexagonally close packed, revealing only occasional, irregular vacancy defects. Spherical caps of the shells were icosahedral containing pentameric units. This relatively simple demonstration challenges the need for spherical viral forms to maintain strictly isometric assembly modes. It is more likely that the coat proteins of spherical viruses are simply programmed to assemble into closed shells with a minimum number of subunit copies. While this number remains constant for the assembly of native particles, with or without the native genome, foreign and rigid templates, the shape of which cannot be changed or condensed, force the assembly to complete with a larger number of subunit copies. Similarly, by capitalizing on the intrinsic polymorphism in the assembly of coat proteins, bespoke virus-like particles can be constructed.[107] CCMV coat proteins disassembled and reassembled around topologically different DNA generate distinctive virus-like topologies: spherical forms are assembled around conformationally flexible single-stranded DNA, whereas rod-like double-stranded DNA templates the assembly of filamentous particles, Figure 3.10b. This property is intriguing in that it is possible to derive empirically the exact numbers of the same protein subunits to close around different shapes with predefined dimensions. Such template-controlled encapsulation properties can find use in different biomedical applications, including targeted drug delivery, diagnostic non-invasive imaging and vaccine development.[107–110] Although unable to replicate outside their native hosts, plant viruses can be repurposed to infect human cells. For example, strategies for the continuous imaging of tissues deep inside living organisms have been pursued using CPMV.[104]

The success of adapting plant viruses for nanoscale designs continues to inspire designs using other viral icosahedra, *e.g.* bacteriophage MS2, turnip yellow mosaic virus and others can equally be used as addressable building blocks for nanoscale designs.[111,112] These proved to be predominantly surface oriented, that is, inner and outer surfaces of viral capsids when altered or redecorated yield novel physical and chemical properties and functions. Filamentous viruses are no exception in this regard. Although tubular forms are not known to be as abundant as spherical capsids, the number of applications of filamentous viruses for nanomaterial synthesis has been significant, notably based on TMV and M13 bacteriophage. Similarly to other viruses, the surfaces of TMV exhibit repeated patterns formed by charged residues, which can serve as nucleation sites for seeding crystal growth, and can be genetically or chemically modified to generate or host specific materials.[113,114] Either side of the virus is suitable for the site-specific deposition of material components *via* chemical modifications or mineralization. For example, platinum complexes reduced at acidic pH precipitated on the external surfaces of TMV capsids, in contrast to photochemically reduced silver(I) salts, which produced aligned nanoparticles loaded in the internal channel of the virus.[115] External particulate platinum was composed of 10 nm particles and encased silver particles were restricted to the diameter of the channel, *i.e.* 4 nm. Nanoparticles aligned inside the virus cavity are well positioned as precursors for metal nanowires as components of molecular electronics. However, the contingency of this concept would depend on a better defined and continuous alignment of particles.

According to first reports, nanowires fabricated by the layer deposition of cobalt and nickel on TMV were a few atoms in diameter, but with lengths extending to hundreds of nanometers to several micrometres (Figure 3.11a).[116] TMV was found to respond well to a number of chemical processes used to template nanotubes of various compositions.[117] The effects appeared to be applicable to other filamentous viruses, such as tobamoviruses, which successfully templated the synthesis of aligned magnetic Pt/Co particles (Figure 3.11b).[118]

The findings also revealed some limitations. The dimensions of aligned nanoparticles and nanowires were restricted by those of the virus core channels, which stimulated the development of approaches to produce similarly

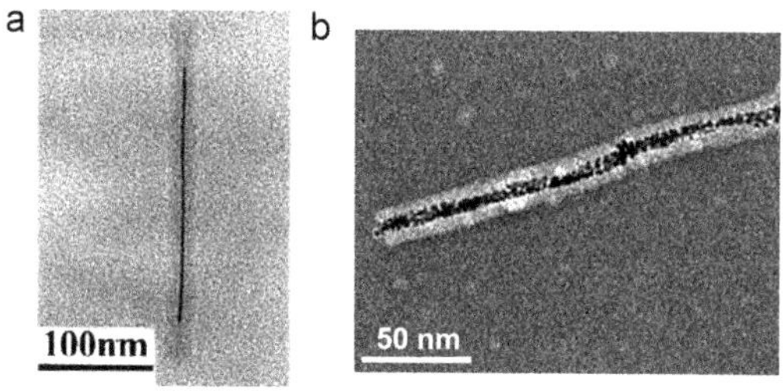

Figure 3.11 Virus-templated nanowires. (a) Electron micrograph of TMV filled with a cobalt wire of over 200 nm in length and 3 nm in diameter. Reproduced from ref. 116 with permission from American Chemical Society, Copyright 2003. (b) Electron micrograph of tobamovirus filled with Pt/Co nanoparticles. Reproduced from ref. 118 with permission from American Chemical Society, Copyright 2010.

discrete but longer structures. Unlike isometric viruses, filamentous viruses having a higher aspect ratio are inherently able to form one-dimensional objects, such as fibrillar nanomaterials, while the propensity for metal-induced aggregation is another limitation to their use in nanomaterial synthesis. This property could be put to use, however. TMV–metal complexes precipitate as a function of virus concentration and pH, particularly when using the divalent ions Cu^{2+}, Cd^{2+} and Zn^{2+}, which mediate TMV precipitation *via* cross-linking.[119] Intriguingly, alkaline earth metal ions (Ca^{2+} and Mg^{2+}) do not, suggesting that metal-binding sites of TMV should be pH responsive.[119]

The anisotropic nature of filamentous viruses was also found to benefit the fabrication of liquid crystalline materials. M13 phage mutants with a specific recognition moiety for ZnS crystal surfaces were shown to self-order into a composite hybrid film ordered at the nanoscale into regular micrometre domains extending over a centimetre length scale.[120] The regular alignment of the nanocrystals in the films suggests that such hybrids can be developed as miniaturized electronic devices, *e.g.* hybrid gold–cobalt oxide nanowires developed on the surface of the phage demonstrated improved battery capacities.[121] Many attempts chose to take advantage of supraviral assemblies. One strategy used highly charged phage particles as polyelectrolyte components in complexation with cationic lipid membranes.[122] M13 phages formed alternately stacked lattices, in which virus layers could afford relatively large pores allowing for packaging nanomaterials, *e.g.* leading to nanoscopic arrays of large (1.2 nm) cations, $Ru(bpy)_3^{2+}$.

Another notable strategy is the use of M13 phage as a building block for self-templating materials. In designing such materials, a comparison is made with naturally occurring systems which self-template their assembly into morphologically defined structures that are not in equilibrium.[123] Self-templating processes depend on and respond to difference kinetic factors, allowing the basic building blocks or units of a given system to assemble into distinctive functional materials. A common feature of self-templating systems is that they adopt supramolecular chirality. This allows the assembly of the same building blocks to generate different liquid-crystal phases underpinning functionally diverse materials. For instance, M13 particles can also be used as suprachiral building blocks. The surface of the phage displays about 2700 α-helical protein subunits with fivefold helical symmetry and a twofold rotation axis, which gives the phage a helical twist at an angle of 41°.[123] Given that the subunits are periodically spaced at 3.3 nm, the twist translates into a supramolecular chiral phase, which can be exploited to build up multiple long-range hierarchical structures *via* a single-step self-templating process. Three different liquid-crystalline structures were demonstrated using this rationale, namely nematic orthogonal twists, cholesteric helical ribbons and smectic helicoidal filaments.[123] Phase transitions and interfacial forces were found to determine the morphology of the self-templated structures. As expected, different material phases gave distinctive optical and photonic properties, which were proposed for use as chiral filters and colour matrices.

Another example is the use of M13 blocks modified with bioactive peptide ligands, such as cell-adhesion motifs, to effectively support cell growth.[123] Moreover, with their high propensity for long-range order, the phage particles could form extended films with conserved and controlled surface patterns. The helical twist in these patterns not only defined the physical characteristics of the resultant materials, but also guided the directional growth of cells which responded to the orientation of the film ultrastructure. Thus, the M13 films provided an extracellular substrate support for cells with specific physical cues, although the role of such cues in cell properties beyond growth and proliferation remains to be understood. Nonetheless, M13 films form a basis for hybrid nanostructured materials such as M13-based matrices incorporating gene delivery capabilities presented by adeno-associated viruses. Such matrices can find use in screening the pathways of targeted gene expression or gene silencing, allowing cross-comparison assessments of gene delivery technologies within the same cellular context.[124] Overall, the tuneability of the physical properties that these virus-derived forms provide underpins functional nanoscale designs which are well beyond known applications in natural systems and can support new functions as diverse and unlikely as piezoelectricity, colour sensors and tissue patterning, Figure 3.12.[124–126]

These developments are encouraging, although in most cases nanomaterials templated on viruses have yet to produce a tangible application and reach out beyond the bench. In addition, only a few viruses have been used to this end. Research in this area appears to progress in small, careful steps, mostly focusing on improving the accuracy and precision of the designs. Designs using other viruses continue to emerge in the literature, but mostly on already beaten paths. The designs will grow in number and the quantity may eventually translate into the quality of predictable structure–function relationships that can be applied to exploitable next-generation nanoscale designs. The potential of the templated synthesis of nanomaterials is significant and may only be fully unlocked by a clear application drive, which

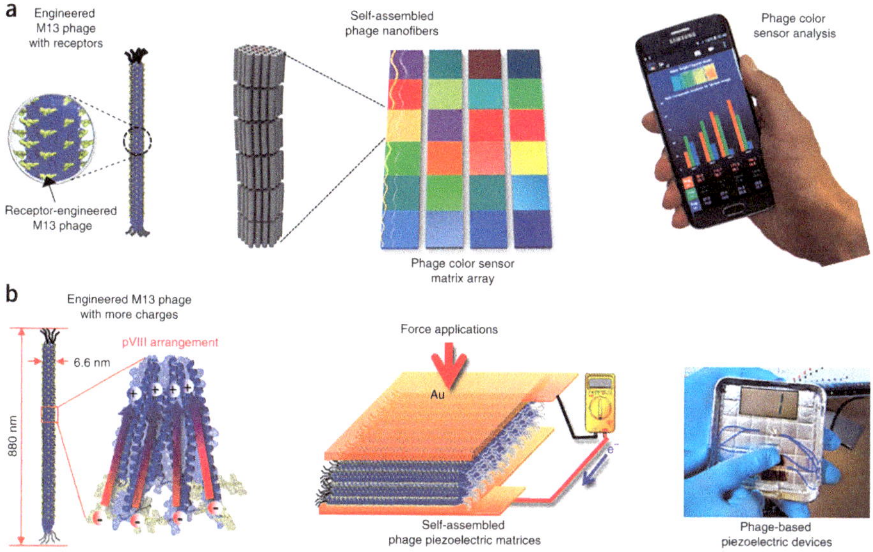

Figure 3.12 M13 phage as a building block for nanomaterial applications. (a) M13-derived colorimetric sensor matrices. A re-engineered phage displays receptors for desired chemicals and assembles into colorimetric thin films that can change colour upon binding target molecules. Through colour analysis and pattern recognition, various molecules can be detected. (b) Phage-based piezoelectric energy-generating matrices exploiting the intrinsic piezoelectric properties of the M13 phage. Upon application of external forces, self-assembled phage matrices can generate electric potentials. The photograph on the right depicts an LCD device that is turned on by tapping on a phage-based piezoelectric device. Reproduced from ref. 125 with permission from Springer Nature, Copyright 2017.

remains to be defined. Other chemistries and more robust manufacturing approaches can be expected to advance the translation of research successes into commercially valuable materials and smart products, including conductive thin-film electrodes, miniaturized batteries, solar cells, light-emitting diodes, sensors, and the like.

3.3 The Challenge of Gene Therapy: Towards an Artificial Virus

Nanoscale designs have been postulated to impact on the development of human society advancing technological progress in the areas of most profound importance. Biological nanosystems, as is evident these days, can offer properties applicable to different sectors of industry. The accuracy, reproducibility and precision with which living systems operate are very attractive for non-biological materials and machinery. However, the prospect of introducing new life-science applications and improving existing technologies provides a more appealing incentive for bionanodesigns. This is of no surprise as our pursuit to decipher biology is relentless, and so is the desire to put living systems to beneficial use. Among life-science technologies where nanoscale designs may provide effective and sustainable solutions are those that are collectively referred to as gene therapy.[127]

It is increasingly apparent that many diseases have a genetic cause. Therefore, gene repair and editing are naturally proposed as an effective means to bring about a treatment or cure. In particular, this concept of genetic therapy is attractive for single-gene disorders such as cystic fibrosis and Duchenne muscular dystrophy, as all it takes to cure these is to correct or replace a defective gene. It is also feasible: there is a growing number of novel therapeutic solutions undergoing clinical trials.[128] In reality, however, the current strategies have persistent shortcomings that prevent the systemic use of gene therapies.[129] In most cases, and fundamentally, unresolved issues revolve around the efficacy and tissue specificity of gene delivery in addition to safety concerns.[130]

Virus-derived systems as delivery vectors continue to dominate the field. Viruses demonstrate performance properties that have yet to be matched by any other gene delivery approaches. After all, these are Nature's forms of choice for effective gene transfer. One can argue, however, that virtually any type of human tissue is a viral host, and that our immune system has developed effective mechanisms to combat viral infections. To be successful, a viral vector must efficiently deliver a therapeutic gene into a cell of interest and evade host defence – the two requirements that have yet to be met. In this light, bionanodesign positions itself as a provider of alternative gene delivery systems. Traditionally referred to as non-viral vectors, such systems are "artificial" and lack the attributes of natural designs. This theme relates to the pros and cons of viral and non-viral vectors and is being revised in favour of what designers aspire towards as an artificial virus. The concept dating back as early as the first considerations of "genetic surgery"[131] is an ambitious attempt at designing a new virus rather than repurposing existing viral forms.[132] With the public's perception towards viruses being increasingly negative, especially given frequent viral pandemics, the use of viral vectors poses yet another setback of public concerns. Proposing artificial viruses may cause even stronger resistance unless and until one can demonstrate benefits that well outbalance the risks. At the time of writing this chapter, an artificial virus is very much an early, fledgling notion of nanoscale structures that resemble viruses in shape, form, physical properties and, with some most recent examples, biofunctional attributes such as infectivity and gene delivery.

One of the core tasks in designing an artificial virus is to establish a structural lead. It should be noted that the design efforts are not necessarily being focused on gene therapy applications. In fact, artificial virus designs meant for therapeutic gene delivery remain in their infancy. A major focus has been on finding structural templates and solutions to the virus architecture in a broad sense, starting with the design of nanoparticles through establishing encapsulating systems and finally the formulation of virus-like structural topologies. Conversely, other designs place a stronger emphasis on finding chemical candidates which are able to deliver nucleic acids into the cell, rather than on the structural consistency and uniformity of the resulting complexes with nucleic acids. Cell traversal and cytoplasmic delivery are among the most critical steps in viral infectivity against which functional mimicry of artificial systems can be judged. These steps occur at a subcellular level and control over them is best realized with nanoscale precision. It is also apparent that nucleic acids, which are nanoscopic objects, may cross the plasma membrane without the intermediary assistance of encapsulating systems. DNA or RNA vectors can be directly injected into skin or muscle tissues, and approaches facilitating this type of delivery are already used for RNA and DNA vaccines. Site-specific injections of viral genes or engineered plasmid DNA encoding viral antigens and inhibitory genes can be introduced, which is advantageous over viral vectors as no anti-DNA antibodies could be observed following such DNA injections. Naked genetic material, that is unprotected by an encapsulating shell, could also be readministered, enabling systemic treatments. However, poor delivery efficacy and low levels of gene expression prompt considerations for delivery vehicles or another auxiliary means.[133] Physical techniques have been proposed to increase the efficacy of targeted delivery using so-called "gene gun" methods, which bombard cells with gold nanoparticles coated with reporter genes.[134] Nanoparticles provide high-density vehicles preconcentrating genes for local delivery into a target cell, but although gene transfer was effective in different cell types, only transient expressions could be observed.

Electroporation is another type of method which promotes the traversal through the plasma membrane, and partly can be used to facilitate delivery into the nucleus.[135] In this approach, high-frequency and low-voltage electric pulses are used to stimulate gene expression, which prompted the term electrogene therapy. Expression levels are 1000-fold higher than for other direct methods and tend to increase the number of cells that express the DNA and the average amount of the gene product.[136] Interestingly, electroporation methods cause little damage to tissues, which encouraged their early translation to successful clinical trials for vaccination and anticancer treatments. Existing evidence suggests that electroporation can be used to target all major tissues of therapeutic relevance, including muscle, liver, lung and vascular tissues. A major limitation of these methods is that the electrodes used to induce electrical fields of appropriate strengths must be of a correct configuration to ensure safe and effective electroporation at the site of injection. Multiple stimulations increasing injection volumes can be applied instead, but even this may not be sufficient to reach the gene delivery efficacies accessible to viral vectors.[136]

A significant part of the problem solved by encapsulating systems is that they protect naked nucleic acids from enzymatic and even mechanical degradation. Thus, gene packaging helps avoid primary opportunistic limitations hampering effective gene expression. Viral vectors are originator viruses encapsulating a genome of interest and hence the design of therapeutic vectors requires recoding of the virus genome, with native capsid scaffolds providing a means for cell targeting and transportation. Nonetheless, viruses have their own deficiencies.[137] These range from insertional mutagenesis, which can trigger carcinogenesis, low gene loading capacities, which are limited to the sizes of native genomes whereas many genetic therapies require the delivery of large DNA and up to the chromosome size, immune responses and undesired interactions with the complement, in addition to the lack of cell and tissue specificity. All these impose a set of additional requirements for an artificial system.

Apart from the major criteria to target and transfect cells as effectively as viruses do, non-viral vectors must be safe and cost-effective to produce, neither toxic nor immunogenic, structurally plastic to encapsulate and transfer genetic material of any size while retaining the discrete form of the assembled shells without aggregation.[138] Nanoscale self-assembling encapsulants or protecting shells wrapping around nucleic acids are the first structural approximations of an artificial virus. Cationic polymers, liposomes and polysomes, amphiphiles, micelles and dendrimers are widely used entities for virus-like designs.[139]

3.3.1 Synthetic Polymer Particles for Gene Encapsulation *Via* Complexation

At its simplest, a virus is a nucleic acid complexed with an oppositely charged polyelectrolyte. All viral genomes, without exception, carry a large negative charge. This simple property is used to design condensation strategies with a polycation, notably polyethylenimine (PEI) and poly-L-lysine (PLL), which readily condense with oppositely charged nucleic acids to form particulate polyplexes. These particles exhibit sizes ranging from 50 to 300 nm, depending on the molecular weight of the polycation used and the ratio of its charge to that of the complexed nucleic acid.[133]

The surface chemistry of polyplexes allows only for a passive route of cellular uptake relying on the buffering properties of the polycations used. Therefore, high cationic density polymers, such as PEI and PLL, are preferred, Figure 3.13.[140]

Protonated forms facilitate efficient gene packaging and transfection and contribute to endosomal escape of the nucleic acid into the cytoplasm. In some cases, complexes may mediate the transfer to the nucleus, although mechanisms of intracellular transfer remain largely unknown. The simplicity of the polyelectrolyte approach, ease of use and construction of polyplexes and their packaging efficiency made these systems popular research tools for *in vitro* applications. However, low transfection efficacies, cytotoxicity characteristics of polycations and polydispersity of the resultant complexes

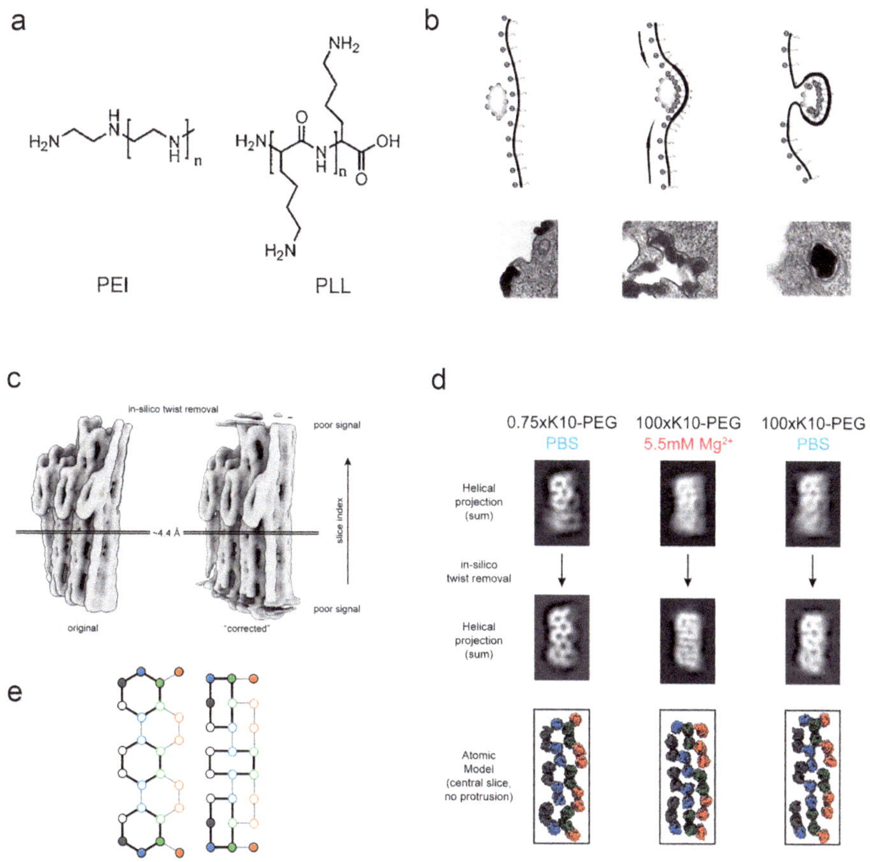

Figure 3.13 Polyelectrolyte version of artificial viruses. (a) Chemical structures of the most common polycations used for gene delivery: polyethylenimine (PEI) and poly-L-lysine (PLL). (b) Schematics and corresponding electron micrographs of the receptor-mediated cellular uptake of complexed DNA showing gradual membrane zipping around cationic particles, which is followed by the formation of size-adjusted intracellular vacuoles. Reproduced from ref. 140 with permission from John Wiley & Sons, Copyright © 2004 John Wiley & Sons, Ltd. (c) Structural changes in DNA caused by polycation coating: original (left) and *in silico* twist corrected (right) cryo-electron microscopy maps of a DNA origami brick. (d) Helical projections of the brick in its original form (top), after *in silico* twist correction (middle) and the central slice of its atomic model (bottom) obtained using different coatings and buffer conditions. (e) Lattice pattern for the regular honeycomb lattice of the original form before coating (left) and compressed honeycomb lattice for DNA coated with poly(ethylene glycol) (PEG)–oligolysine (right). Reproduced from ref. 147 with permission from American Chemical Society, Copyright 2021.

are among the main factors requiring improvements. Instead of full-length, high molecular weight polymers, short oligomeric blocks modified at the termini with cysteine residues can be used. These blocks are equally effective in condensing nucleic acids, but also allow oxidative polycondensation, which in turn can be used to facilitate the disassembly of polyplexes in the reducing environment of the cell.[141] A main improvement of using such a pH-controlled assembly is that it ensures significant decreases in cytotoxicity compared with those of PLL. This relatively simple design also renders cationic polyplexes responsive to the intracellular environment, ensuring 100-fold increases in gene expression compared with those of conventional PLL vectors. A poor capacity for endosomal escape is another recurring issue for these polyplexes. For cysteine-containing short peptide blocks this shortcoming may be attributed to cysteine residues being inert residues and therefore their inclusion in the sequence decreases the buffering capacity of the polyplexes. The incorporation of pH-sensitive and weakly cationic histidine residues is a good alternative to cysteines in promoting the endosomal escape and the disassembly of polyplexes in the cell.[142]

Lysine, cysteine and histidine residues can support functional modifications of polyplexes, including the coupling of molecular ligands recognizing specific cellular receptors. Multiple modifications may also be beneficial. The performance of polyplexes appears to correlate with their chemical topology, *e.g.* linear or branched, and their size. Cross-comparisons performed for linear, dendritic and hyperbranched polycations revealed that branching facilitates enhanced transfections, but also causes more pronounced cytotoxic effects.[143] To obtain correlations in biological activities between different topologies requires the consideration of other factors, such as the molecular weight of the polycations used. Linear PLLs give much higher cytotoxicity levels compared with those of their dendritic and hyperbranched analogues. Low molecular weight polycations induce immediate cell death *via* osmotic shock, which is in contrast to higher molecular weight polycations disrupting cellular membranes on contact. Apoptosis proved to be more apparent for branched polycations than linear analogues of the same molecular weight.[139] This can be attributed to differences in biodegradability. Branched polycations containing isopeptide bonds are only partially biodegradable. Similar trends are observed for hybrid polycations such as histidine–lysine polymers. Larger polymers tend to provide more effective packaging and protection from enzymatic degradation. However, these are also prone to cause higher levels of toxicity. Achieving the right balance between these properties is of particular importance for *in vivo* applications. New formulations of polyplexes and their hybrids appear to give safe and more effective vectors for gene transfer *via* different injection routes, including intratracheal instillation, kidney perfusion and intravenous injections. However, the biocompatibility and morphological stability of these vectors limit their use beyond experimental medicine, stimulating the exploration of naturally occurring biopolymers.

Ubiquitous to various organisms are gelatine and chitosan. These are enzymatically more stable than PEI and PLL and are less toxic. These polymers readily complex with DNA to form nanoparticles, which can be delivered into target tissues *via* routes atypical of synthetic polycations. Engineering graft hybrids of natural and synthetic polycations is also considered, mainly in attempts to improve physicochemical properties compared with those of the constituent components and to generate more stable and less toxic materials. Hybrid polyplexes appear to exhibit higher buffering capacities than naturally occurring biopolymers, and which can be further improved *via* PEGylation, which is used to derivatize macromolecular systems with poly(ethylene glycol) (PEG) molecules of various lengths. PEG is a hydrophilic, but inert, neutral polymer, which allows it to tune the physical properties of polyplexes and other vectors, including viral systems.[144–146] PEG brushes can passivate vector surfaces, reducing cytotoxicity and improving biocompatibility. Polycationic coatings and their PEG hybrids continue to be used to stabilize and protect nucleic acids, which is important for media with low ionic strength. It is also becoming apparent that polycation coatings may have an effect on the structure of nucleic acids with which they complex. Twisting and bending are among the structural deformations observed in DNA. For example, DNA origami test models designed as honeycomb helical lattices were deformed into square lattice-like patterns by such coatings.[147] The coated DNA tended to shrink along the helical axis, expanding in the direction perpendicular to it, but also withstood the low ionic strength of physiological media. Without the coatings, the model disassembled in the media. The results confirmed polycationic coatings as an effective protection for DNA and supramolecular DNA structures, *i.e.* stabilizing non-covalent forms, and showed also that the coating itself remained stable in these complexes.[147]

The relative simplicity of the preparation and polyelectrolyte properties of polycations make them amendable to empirical improvements. Polydispersity at the primary (polymeric chains) and complexation levels limits the polyelectrolyte version of an artificial virus to a very broad range of irregular forms the properties of which are subject to external stimuli, that is, the conditions of preparation and usage. The responsiveness of polycationic systems to environmental changes is exploited to control the size distribution of the resultant polyplexes: polycations mixed with nucleic acids rapidly give smaller complexes, whereas thermodynamically sustained titrations give larger and more amorphous structures. These properties manifest the lack of control over polyplex designs, including structural uniformity, consistency and stoichiometry and the cargo load of polyplexes. Effective packaging and shielding properties of polycations aid in the delivery capabilities of polyplexes, but poor surface chemistry strips the polyplexes of biofunctional recruitment, limiting delivery routes to buffering effects. Other polymers which can offer richer chemistry and a better defined propensity for hierarchical assembly can offer complementary solutions.

3.3.2 Cationic Peptides Condensing Genes for Intracellular Delivery

DNA in the cells or viral genomes in capsids are associated with cationic residues. More detailed studies revealed that the residues compile conserved motifs of short peptide sequences (<20 amino acids). The chemical synthesis of peptides of this size range is well established and relatively straightforward, with a number of naturally occurring and synthetic peptide sequences already having been proposed as gene condensing agents. These range from C-terminal cationic segments derived from virus core proteins, *e.g.* from adenovirus or hepatitis viruses, to the peptide repeats of histones and histone-resembling protein sequences. Synthetic peptides based on these and similar motifs wrap or condense nucleic acids, mediating the intracellular uptake of the complexes formed in a receptor-independent manner.[148,149]

Typically, these are arginine-rich sequences exhibiting also appreciable cell-penetrating properties. Charge ratios between the target nucleic acids and peptides define the physical properties of the complexes formed. Complexation also tends to occur as a hierarchical stepwise process, especially when involving other components. For example, peptide–DNA complexes in the presence of cationic lipids produce virus-sized nanoparticles. These particles are solid and stable to aggregation and allow an informed choice to be made of components with specific properties, such as an enhanced ability for endosomal escape. The composition of the nanoparticles can be tuned to promote the synergistic enhancement of transfection *in vitro* and *in vivo*. A caveat in designing these systems is to ensure that the components are incorporated in ratios enabling synergistic assembly and disassembly, which is particularly useful for endosomal escape properties.[150] The latter remains a limiting factor for artificial gene transfer systems.

Without an effective means of endosomal egress, genes of interest cannot be released into the cytoplasm and expressed. The importance of this property for gene delivery systems can be demonstrated using lysosomotropic agents. A common agent is the antimalarial drug chloroquine, which was shown to enhance transfection considerably in different non-viral vectors.[151] Although the exact mechanism of its action remains to be elucidated, chloroquine is widely accepted as an endosome-release reagent in gene transfer experiments. Cationic membrane-active peptides, which are found in virtually all organisms as effector molecules of innate immune systems, are popular alternatives. Both naturally occurring and synthetic sequences are being used to construct gene delivery systems. Most of these peptides exhibit cell-penetrating properties by virtue of being cationic. In this regard, they are similar to cell-penetrating peptides – cationic sequences that effectively traverse the plasma membrane and have been shown to mediate the delivery of various cargo types.[152] These peptides are prototypical arginine-rich sequences (<10 amino acids) of protein transduction domains, including the antennapedia homeodomain of *Drosophila*, the transactivator of transcription (TAT) protein identified in HIV, DNA-binding protein VP22 from herpes simplex virus, the neuropeptides galanin and mastoparan and many others.[153–155]

Cell-penetrating peptides share structural similarities with other membrane-active peptides, suggesting a common mode of interactions with membranes, which could prove instrumental for emulating processes that viruses employ to disrupt endosomes. Inspiration can come from endosomal acidification, which triggers conformational changes of viral proteins initiating the lysis of endosomal membranes.[156] For example, PLL vectors conjugated with the N-terminal sequence of haemagglutinin HA-2 from influenza virus enhance gene transfer probed in different cell lines.[157] Analogously, combinations of cationic and anionic peptides can be used to condense DNA, cationic peptide, and induce a pH-dependent endosomal release of the DNA, anionic peptide. Such a vector can be constructed by mixing DNA with the cationic peptide component, resulting in a positively charged polyplex, followed by the addition of the anionic component, which activates the vector for endosomal release, leading to high gene expression levels.[158]

Covalent nucleic acid or oligonucleotide conjugates with cell-penetrating peptides are used as alternative transfection tools to particulate forms of non-viral vectors. Encouraging results have been reported using these systems both *in vitro* and *in vivo*, achieving the levels of systemic use towards corrective therapy applications. In particular, peptide conjugates with splice-switching oligonucleotides proved to advance solutions for neuromuscular diseases such as spinal muscular atrophy. Profound phenotypic corrections in peripheral and central nervous systems were reported for such conjugates at doses an order of magnitude lower than required by naked oligonucleotides.[159] Despite the clinical potential of covalent conjugates, their application is limited to the delivery of short oligonucleotides. There is also scarce information on whether these conjugates are prone to assembly or aggregation as could be expected for oppositely charged components. The lack of data in this regard can be attributed to the fact that oligonucleotides used in these conjugates are predominantly neutral analogues of single-stranded DNA, *e.g.* phosphorodiamidate morpholino oligomers.

Localized targeting or receptor-mediated targeting is another advantage that is intrinsic to peptide-based gene delivery systems. Specific cell targeting may not be necessary for *in vitro* studies but is critical for the cellular uptake of polyplexes *in vivo*. Acquiring this property almost negates the role of the buffering capacities of polyplexes in mediating their cell entry. Cell-adhesion or homing peptide motifs incorporated into a non-viral vector target it to a cellular receptor, which presents a simple and appealing strategy for cell targeting. However, known motifs target receptors present in different cell lines, and none is unique to a particular cell or tissue type, which makes this approach little different from the non-specific cell targeting by viruses. Peptide motifs such as RGD of fibronectin or YIGSR of laminin target integrins and integrin-associated receptors that are ubiquitous to many cell lines, and their incorporation into gene delivery vectors can significantly improve the efficacy of gene transfer.[160,161] Nuclear localization signal sequences constitute another class of homing domains, short peptides that can transfer genetic materials into the nucleus, enabling the expression of therapeutic proteins.[162] Using these motifs and their combinations, a number of cell-targeting strategies have been proposed, with most being focused on gene delivery to tumours and vascular tissues in which cell receptors are overexpressed and therefore are more likely to be targeted by the vectors than other cells. The same strategies are applicable to other classes of non-viral vectors, the chemistries of which are compatible with peptides, *e.g.* lipid-based systems.

3.3.3 Lipid, Glyco and Peptide Amphiphiles for Vesicular Nanodesigns

Lipid- or liposome-based gene delivery systems have been extensively used since they were first described in the early 1960s,[163] and with the subsequent introduction of the lipofection DNA transfer procedure,[164] the principles of which remain in use to this day.

Liposomes are spherical lipid bilayers that close around an aqueous cavity, Figure 3.14. Upon complexation with nucleic acids, liposomes become

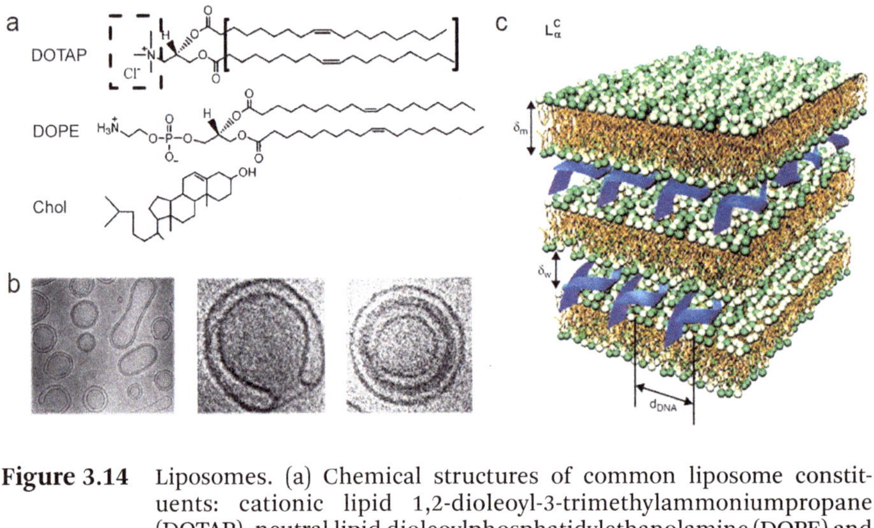

Figure 3.14 Liposomes. (a) Chemical structures of common liposome constituents: cationic lipid 1,2-dioleoyl-3-trimethylammoniumpropane (DOTAP), neutral lipid dioleoylphosphatidylethanolamine (DOPE) and cholesterol (Chol). The polar head and hydrophobic tail of DOTAP are indicated by a dashed rectangle and square brackets, respectively. (b) Electron micrographs of DOTAP:Chol liposomes without DNA (left and middle) and with DNA packed between two bilayers of DOTAP:Chol (right). Reproduced from ref. 165 with permission from Springer Nature, Copyright 1997. (c) Schematic representation of the lamellar phase of cationic liposome–DNA complexes (L_α^C) with alternating lipid bilayers and DNA monolayers. The DNA interaxial and interlayer spacings are denoted d_{DNA} and $d = \delta_w + \delta_m$, respectively. Reproduced from ref. 167 with permission from Elsevier, Copyright 2001.

lipoplexes. Both cationic and anionic formulations can be used. Anionic liposomes are less popular owing to their inefficient encapsulation of nucleic acids, whereas cationic lipids readily interact with nucleic acids or oligonucleotides to form vesicular aggregates, which convert into lipoplexes upon reaching a density threshold of lipids. Similarly to polyplexes, cationic lipoplexes easily become fused with plasma membranes, but appear to escape endosome more effectively. Complexation mechanisms are also somewhat similar – lipid molecules bind along the helices of nucleic acids rather than encapsulate them as viruses do, Figure 3.14b and c.[165–167] The size and topology of the complexed nucleic acid impact on the form and size of the resulting lipoplex particles, although their size distribution is fairly broad, ranging from 20 nm to several microns in diameter, Figure 3.14b.[165] The likely reason for this polydispersity lies in the nature of the liposomal architecture and components.

A cationic lipid is an amphiphilic molecule comprising an aliphatic chain and a cationic headgroup. These are primary amphiphiles that self-assemble in water largely *via* the hydrophobic interactions of lipid aliphatic tails. The molecular parameters of amphiphiles such as hydrophobic volume, the length of the aliphatic chains and the headgroup areas determine not only the final morphology of a lipoplex, but also the self-assembly pathway that it takes, of course in conjunction with the nucleic acid with which it complexes. The surfactant concept is normally used to describe and predict the impact of these parameters on lipoplex self-assembly.[168] A surfactant parameter S, defined as

$$S = v/la_0$$

where v is the hydrophobic volume specified by the hydrocarbon tail, l is the length of the tail and a_0 is the effective area per headgroup, defines geometrical constraints of the lipid packing. Specifically, S relates the properties of lipid amphiphiles with the curvatures of preferred assemblies, which can be positive and negative. Assemblies with positive curvatures are curved around their hydrophobic part (normal phases), whereas negative curvatures favour polar groups (inverted phases). Therefore, a simple rule can be derived to predict optimal assemblies in water, with more curved and contained assemblies having smaller S values: $S < 1/3$, $S = 1/3–1/2$ and $S = 1/2–1$ for spherical, cylindrical and lamellar assemblies, respectively. Amphiphiles with two aliphatic tails would have larger S, giving rise to lamellar structures, whereas more polar amphiphiles would assemble into spherical structures the shapes of which are less curved with increasing amphiphile concentration.[168]

Nonetheless, the usefulness of the surfactant theory for designing lipoplexes is limited. Liposomes are composed of concentric membranes each 4 nm thick.[166] The number of such membranes can vary within one liposome type, and which the surfactant theory cannot predict, while suitable, unifying models are required to design artificial virus-like systems based on lipid assembly. Cationic lipoplexes remain in use and often show superior gene transfer capabilities to those of polycations or peptides.

Liposome-based systems also dominate the representation of non-viral vectors in clinical trials, showing promising results in the treatment of cancers and cystic fibrosis, whereas commercially available lipoplex formulations are popular research tools for transfection studies *in vitro* and *in vivo*.[169] Lipoplexes based on quaternary ammonium amphiphiles such as Lipofectin, LipofectAmine, Cellfectin, Transfectam, LipoTaxi and TransFast demonstrate high transfection efficacies, although mechanistic factors impacting on the consistency of these results remain poorly understood. The most recent studies provide evidence linking the ability of lipid-based reagents to fuse and disrupt plasma and nuclear membranes with cytotoxic effects in a manner similar to that observed for polycationic vectors, and the stoichiometry of nucleic acid and lipid amphiphiles, with both components responsive to external stimuli, appears to impact on the morphology of the final assemblies.[170–172]

As an exemplar, lipidic amines such as spermine and spermidine can condense DNA into toroids and rods of size 70–100 nm.[173] These structures are unstable under physiological conditions. The hydrophobic tails aggregate and, by virtue of being physically linked to DNA-binding cationic headgroups, stabilize the packing of nucleolipidic nanoparticles. The resultant nanoparticles vary in size depending on the amphiphile-to-DNA ratio and undergo relatively rapid decomposition in the bloodstream. In liposomes, nucleic acids can become incorporated in the aqueous core and in the lipid bilayer of amphiphiles. As a consequence, the number of nucleic acid copies per lipoplex can vary, unlike in viruses hosting one copy of a capsid.

According to the original "beads-on-a-string" model for lipoplexes, several liposomes bind along the DNA strand. The model is based on electron microscopy observations of different structures, including plasmid-sized strings and oligolamellar and tubular structures. Another model based on the data from *in situ* microscopy and X-ray diffraction analysis considers that DNA induces a topological transition of liposomes into collapsed condensates of 1 μm globules.[174] The globules (lipoplexes) occur as a result of DNA screening electrostatic interactions between lipid bilayers, leading to the formation of higher order multilayers.[166] DNA is said to be sandwiched between the planar arrays of the lipid bilayers. The 6.5 nm thick arrays appeared highly periodic: a 4 nm thickness of a lipid bilayer separated DNA helices in the plane of a sandwich, leaving the remainder of 2.5 nm to the aqueous cavity hosting DNA. This agrees with the encapsulation of a single layer of hydrated DNA.

The model gives a good description of the liposomal encapsulation of DNA, but to predict and rationally control the structure of lipoplexes requires more empirical considerations of how the structure of the components links with the sizes of lipoplexes. One factor to consider is the weight ratio of lipid to DNA.[167] As more DNA is added, globular lipoplexes, coexisting with liposomes used in excess, fuse and enlarge, indicating that lipoplex formation is intrinsically polydisperse. Similarly to polycations and unlike peptides, lipids lack a means of supporting specific molecular recognition and do not adopt well-defined secondary structures, which could otherwise introduce structural constraints promoting more discrete and uniformly sized assemblies. Therefore, lipid-based systems are subject to the continuous pursuit of optimized designs and formulations. As described earlier, the use of short oligomeric blocks, instead of full-length polymeric cations, for polyplex formation allows certain improvements to be achieved. Lipid amphiphiles used in lipoplexes are already short blocks. However, simple aliphatic systems modified with elementary biomolecular units, such as amino acids or sugars, show further improvements regarding the monomolecular condensation of DNA. Small cationic detergents containing thiols can complex DNA at concentrations below the critical micelle-forming concentration (cmc). Mononuclear lipoplexes are formed in two steps: in the first step, the lipid detergent has a high cmc, sufficient to initiate DNA binding and condensation, but complexation is completed with lipoplex formation at a very low cmc and is further stabilized by thiol oxidation. This helps prevent cationic micelle-mediated aggregation of DNA particles, giving rise to discrete 30 nm complexes, which are stable enough to be transferred to the cell nucleus, Figure 3.15a.[175]

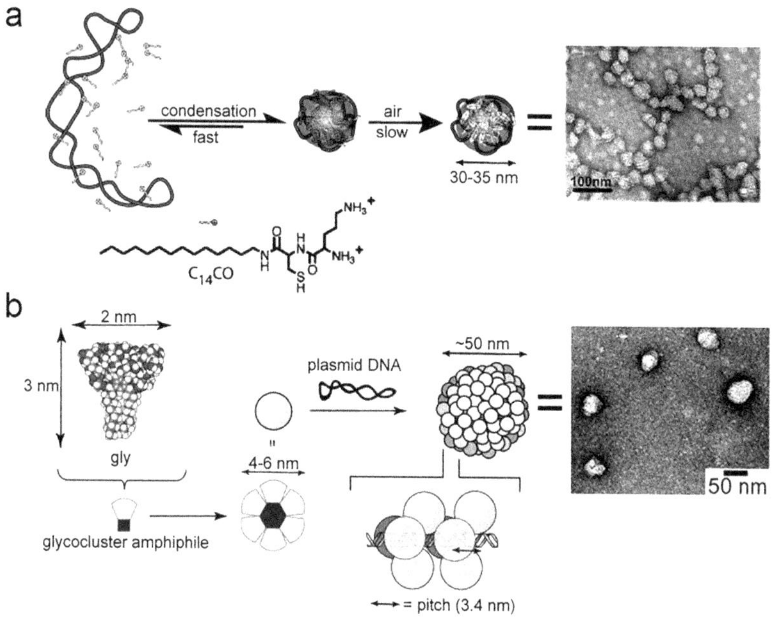

Figure 3.15 Self-assembled amphiphiles. Schematics of DNA complexation by (a) cationic thiol-containing and (b) neutral glycocluster amphiphiles leading to the formation of (a) monomolecular DNA nanoparticles and (b) "glycoviruses". Corresponding electron micrographs are also shown. Adapted from ref. 175 and 176 with permission from American Chemical Society, Copyright 2005, 2003.

Questions remain, however, regarding the ability of these and other lipoplexes to undergo decomplexation in the cell. Poor stability of gene delivery vectors of this type *in vivo* exposes common drawbacks of artificial systems.

An original alternative was proposed in the design of glycoconjugate amphiphiles assembling into glycoclusters.[176] These are regular pentameric virus-like micelles, which prompted the term glycoviruses, also reflecting the particles' mononuclear and stoichiometric assemblies, Figure 3.15b. This design is different from that of systems based on cationic lipids or polymers that cause nucleic acids to collapse into viral-sized complexes. The glycoclusters are strongly adhesive to nucleic acids but not to each other, which allows them to coat genes in a way similar to that of viral capsomers. Glycoparticles bind to the major groove of DNA. The groove spans 1.3 nm and has a 3.4 nm pitch, suggesting that four 4 nm glycoparticles can be shared between two pitches, which was sufficient for the fully formed glycovirus assemblies to exhibit transfection efficiencies an order of magnitude higher than those of Lipofectin.[176] This is important as such neutral structures have no means for endosomal buffering displayed by cationic systems. In this light, mechanisms that accept cation-mediated endosomal release as a main determinant for successful gene transfer are not applicable to the glycoviruses and may not be a prerequisite for artificial viruses in general. The fully formed glycoviruses were compactly packed, charge neutral and were taken up by cells *via* a pinocytic receptor-specific pathway.[177] The concept opens up promising avenues for further nanoscale designs, particularly with respect to non-cationic virus-like shells whose p*I* values are within the range of the isoelectric points of naturally occurring viruses (3.5–7).[178] Small neutral or anionic virions are anticipated to cross the blood–brain barrier more effectively than cationic particles. However, irrespective of the nature of the particles used, a universal set of design parameters that can mitigate the main challenges towards tangible gene delivery to match the efficacy of viral vectors has yet to be demonstrated. Much research is still required to identify better and more quantitative approaches for the complexation of nucleic acids using the outlined chemical classes. Emphasis here is increasingly placed on *in vivo* tests. The success of gene delivery *in vivo* considers more factors and depends on the route of administration and on the tissue type targeted. The impact of *in vivo* environments on the physicochemical properties of gene delivery formulations is also much greater than that of *in vitro* validations. Transfection efficiencies can vary by a factor of 1000, with many tissues requiring specific and discriminative targeting. Therefore, strategies focusing on condensation of nucleic acids as opposed to encapsulation are nowadays aimed at finding optimal formulations for systemic use *in vivo* and ultimately in clinic.

Another body of research is focused on designing encapsulating systems mimicking the physical properties of viruses. The main criterion in this research is the formation of hollow vesicular assemblies. Typical designs assemble from di-block and facile amphiphiles: peptides containing two explicit faces, polar and hydrophobic, corresponding to two sequence segments or two sides of a constrained framework such as a cyclopeptide.

For instance, a di-block copolypeptide amphiphile can be made of a cationic polyarginine block and a hydrophobic polyleucine block, Figure 3.16a.[179] As cell-penetrating sequences, poly- and oligoarginines are active promoters of intracellular uptake, whereas leucine residues have a high propensity for helix formation. In water, the hydrophobic helical segments cluster together in an antiparallel manner forming outer and inner arginine layers of vesicular particles. Cyclic peptides with alternating hydrophobic and cationic residues assemble into similar vesicles, with hydrophobic residues clustering into one hydrophobic interface, which separates two polar surfaces of the vesicles, Figure 3.16b.[180]

The cationic surfaces of these and other analogous designs self-promote the effective transport of the resultant vesicles across the cell membranes. However, the size distributions of these structures are broad, and although they can be controlled by extrusion after the assembly, the need for size separation indicates the main shortcoming of these vesicular types, namely polydispersity in the assembly, which may result in aggregation or fibrillation.[180] These drawbacks re-emphasize the promise and technological potential of symmetry designs, which constitutes a major research endeavour aimed at creating symmetrical protein cages. The exploitation focus of this research is to provide design principles building upon the rules of biomolecular self-assembly and to apply such principles to engineer synthetic virus-like architectures *de novo*. In one way or another, known attempts in this

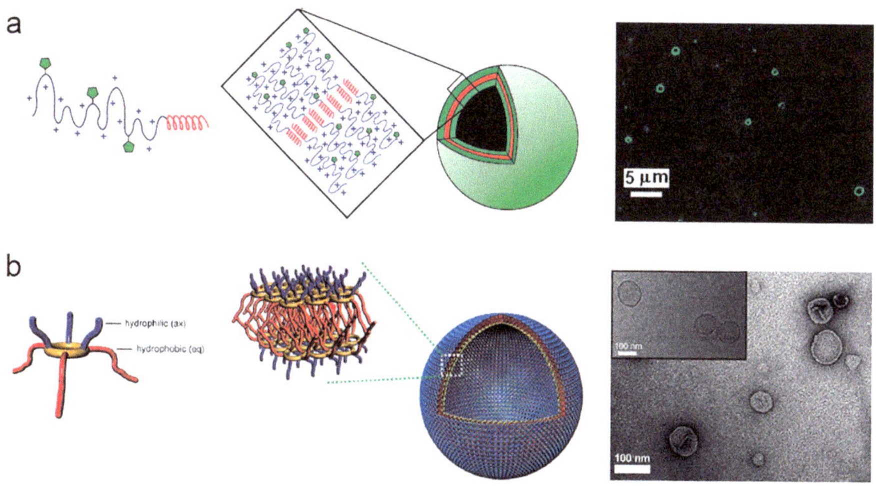

Figure 3.16 Self-assembled facile amphiphiles. (a) Schematics and micrograph of vesicular assemblies from a di-block copolypeptide amphiphile. Reproduced from ref. 179 with permission from Springer Nature, Copyright 2006. (b) Schematics and micrograph of vesicular assemblies from a cyclic peptide facial amphiphile. Reproduced from ref. 180 with permission from John Wiley & Sons, Copyright © 2010 WILEY-VCH Verlag GmbH & Co. KGaA, Weinheim.

direction strived to design regular, symmetry-defined structures. Designs that consider nucleic acid encapsulation are in their infancy, although several impactful systems and topologies have been reported and some examples have demonstrated potential for gene transfer *in vitro*.

3.4 Regular Geometric Forms for Virus-like Designs

One of the early attempts towards regular cage-like designs was protein nanohedra.[181] This approach chose to match symmetry-driven assembly with the polyvalence of protein building blocks. The design uses a monomeric protein block built of protein domains each supporting a distinctive oligomeric state. Two domains, dimeric and trimeric, are conjugated as one asymmetric protein block. Each of the two independent parts binds to its counterpart of another copy of the protein, thereby allowing the assembly of symmetric nanohedra. The way in which domains are arranged in the protein block in relation to two virtual symmetry axes predetermines the final assembly form, which is enforced by a rigid linker between the domains. Twelve copies of this chimeric protein were estimated to follow a cubic point symmetry to assemble into a small (18 nm) tetrahedral cage with an edge length of 15 nm. Consistent with the design, experimentally discrete particles with diameters of 12–15 nm were observed, with particles with sizes outside this range also being present, suggesting polymorphism in assembly.[181]

Following on from this first design, which used naturally occurring dimeric and trimeric proteins, the concept was extended to a set of geometric rules proposed to construct cubically symmetric cages.[182] It was shown that two oligomeric protein components representing two simple symmetries can be joined up into a geometrically defined arrangement. Each of the components is fused to a coiled-coil strand, which is complementary to the coiled-coil strand of another component. When mixed together, the strands form a rod-like, rigid heterodimeric super-helix, which connects the two protein components. Because the components support different oligomeric states and the super-helix limits flexibility, geometrically unique structures can be created. For example, dimer–trimer and trimer–tetramer combinations generate tetra- and octahedral structures, respectively, Figure 3.17a.

Analogous designs set out by adopting icosahedral symmetry for regular polyhedra from complementary coiled-coil polypeptides. In this design, a polyhedron was predicted to be built from multiple copies of one asymmetric unit, while the least common multiple of different oligomerization domains comprising the unit defined the type of polyhedron assembled, Figure 3.17b. The asymmetric units constructed are 3D tripyramids, with each of their edges occupying one of the rotational symmetry axes, hence oligomerization domains can be superimposed onto the edges and when fused would yield a 3D self-assembling block, Figure 3.17b.[183] Indeed, a contiguous polypeptide sequence comprising pentameric and trimeric α-helical coiled-coil domains,

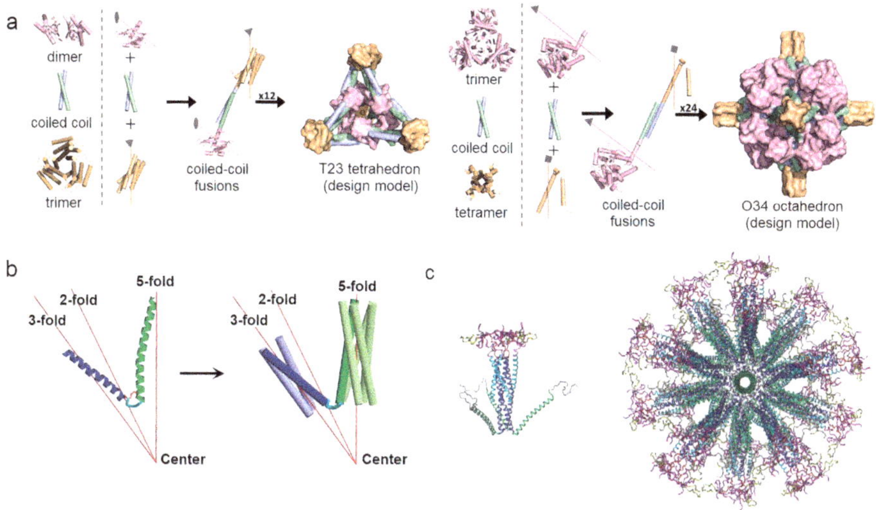

Figure 3.17 Self-assembled nanohedra. (a) Schematic of two symmetrical design models each comprising two distinct oligomeric domains fused together by a heterodimeric coiled-coil linker (green and blue). Left panel: a tetrahedral design is assembled from a C_2 dimer (pink) and a C_3 trimer (orange). Right panel: an octahedral design is assembled from a C_3 trimer (pink) and a C_4 tetramer (orange). Surface representations of the designs are shown (right). An oval, a triangle and a square denote C_2, C_3 and C_4 symmetry axes, respectively. Reproduced from ref. 182 with permission from American Chemical Society, Copyright 2021. (b) A 3D building block consisting of two coiled-coil domains with threefold (blue) and fivefold (green) symmetries separated by a short linker (cyan). The intrahelical disulfide bridge bending the block is shown in red. Thin red lines indicate rotational symmetry axes. Reproduced from ref. 183 with permission from Elsevier, Copyright 2006. (c) Three monomers assembled as a trimer (left) and 3D nanoparticles assembled from 20 copies of the trimer (right). HIV antigens are shown in red, magenta and cyan. Reproduced from ref. 184 with permission from Elsevier, Copyright 2019.

to arrange five- and threefold symmetry axes of the polyhedron, was shown to assemble into a regular polyhedron. The assembly was complete in two steps: 15 copies of the block first arranged into an "even unit", followed by the association of even units into larger and closed structures. The morphology and size of resulting nanoparticles correlate with how regular the geometry of the units is, permitting some level of structural polymorphism. Indeed, spherical nanoparticles with variable diameters in the range 15–45 nm were found. Careful adjustments to peptide refolding proved to aid in the generation of more uniform nanoparticles, re-emphasizing the role of folding in self-assembly. These designs are unique in that they are not cage- or shell-like structures, but spherical nanoparticles resulting from a brush-like arrangement of the coiled-coil blocks converging towards the core, which is reminiscent of covalent dendrimer structures. Such an arrangement lends itself as a suitable platform to display simple and complex antigens – the approach that has found an experimental proof for engineering synthetic vaccines displaying HIV antigens in native-like conformations, Figure 3.17c.[184] The use of sequences based on ubiquitous secondary structure motifs as autonomous units for nanoscale designs is another valuable demonstration of the outlined designs of spherical and shell-like assemblies. This strategy creates an ample opportunity for employing amino acid sequences of *de novo* design, without the need for adapting wild-type proteins.

Among other structural motifs, coiled coils appear to be the most straightforward for the design of closed shells and particles. A number of polypeptide cage designs have been demonstrated, including those that adopted molecular topologies, which have no natural analogues, *e.g.* dendrimers, but capitalize on natural phenomena, *e.g.* shell-like assemblies, to introduce an application level or new function. This can be demonstrated using a polynanocage assembly.[185] Unlike protein shells, which host one cavity, this assembly was designed to comprise multiple cavities, which allows it to encapsulate multiple material copies. Thus, it can be said that an individual polynanocage of this type contains multiple shells, which renders it porous, Figure 3.18a. This design is unique in that each cavity it hosts is structurally and functionally independent. Therefore, it required a suitable molecular topology, but also simple such that it can accommodate a coiled-coil motif. The dendrimer architecture offered an optimal solution. A non-covalent dendrimer framework was assembled from a single peptide exhibiting two polar faces, designed to propagate into a network of coiled-coil dimers. Nanoscale cavities in this network occupied sixfold symmetry vertices. To increase their size, another subunit, complementary to the first peptide, was introduced. The two monomeric components formed a spherical, shell-like assembly, the ultrastructure of which was reminiscent of a non-covalent, supramolecular dendrimer, Figure 3.18a.

Dendrimers are symmetrical molecular topologies characterized by multiple branches radiating from a central core.[186–188] This endows them with unique properties, including a high capacity to display molecules on their exterior, which can be exploited in the design of vaccines, and the formation of multiple cavities or cores as confined environments for the synthesis of nanomaterials. Multiple, symmetric and regular branching is the hallmark of the dendrimer architecture.[189] Branching cells in the covalent dendrimer framework are potential cavities if made hollow, as realized in the polynanocage design. Supramolecular dendrimers thus present obvious scaffolds for nanoscale cages with demonstrable potential for gene delivery and molecular encapsulation.[190,191] In relation to the synthesis of nanoscale materials, polyvalency or multiple cages can amplify the yield of a given nanomaterial per individual assembly multifold. As an exemplar, the reduction of ionic to colloidal silver in designed polynanocages generated extended spreads of uniformly sized silver nanoparticles, Figure 3.18a.[185] The shape and size of the spreads were nearly identical with those of the polynanocages and their cavities, where the particles were produced.

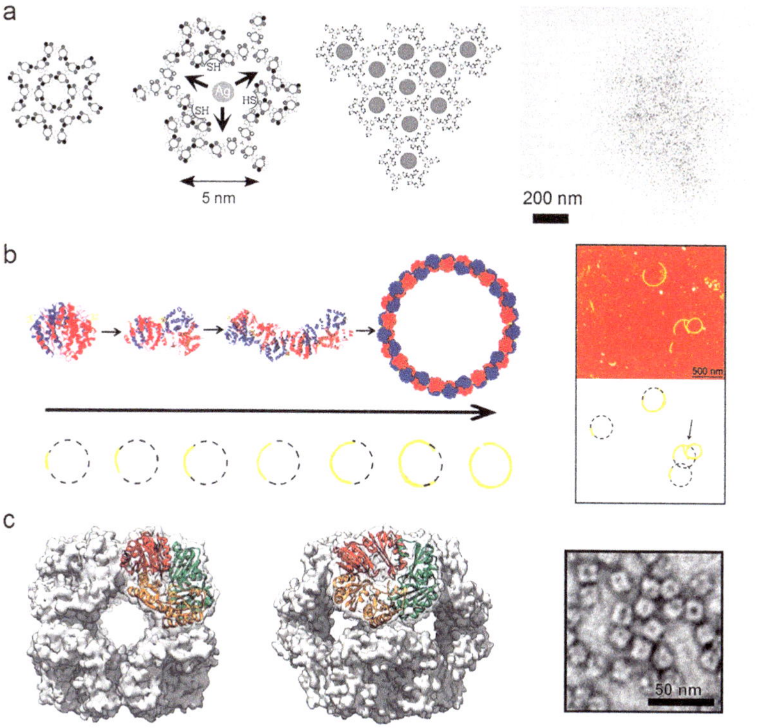

Figure 3.18 Self-assembled ring-like nanocages. (a) Schematic of a polynanoreactor design showing a single cage (left) and when extended (middle) to host a silver nanoparticle (right). Far right: an electron micrograph of a spherical spread of silver nanoparticles synthesized within an individual reactor, which was subsequently degraded enzymatically. Reproduced from ref. 185 with permission from John Wiley & Sons, Copyright © 2007 WILEY-VCH Verlag GmbH & Co. KGaA, Weinheim. (b) Left: schematic representation of the formation of protein nanorings through metal (Ni^{2+}) coordination. As protein monomers come together, the assembly exhibits a bending trend, eventually resulting in a nanoring. Right: an atomic force micrograph (top) and the curvature of the protein half-rings in the micrograph followed schematically (bottom). Arrows indicate overlap and twisting of two collided "growing arcs". Reproduced from ref. 193 with permission from American Chemical Society, Copyright 2013. (c) Surface models of a self-assembling cage viewed along fourfold and twofold symmetry axes, with one trimer shown as a coloured ribbon (left), and an electron micrograph of assembled rings (right). Reproduced from ref. 196 with permission from American Chemical Society, Copyright 2018.

Indeed, ultrastructural analysis of individual polynanocages revealed that cavities were fused into a network of 5 nm rings, which templated the synthesis of 5 nm silver particles. Each ring in the assembly is supported by other rings connected to it in the fusion. Thus, the assembly of the rings in the polynanocage requires high cooperativity, without which the assembly of individual cages could not take place.

However, discrete nanoscale ring-like assemblies are also possible.[192] It has been shown that wild-type proteins tethered together by a flexible linker can be assembled into stable ring-like structures 8–20 nm in diameter, suggesting that each ring is a discrete assembly of 4–8 dimerized proteins. This type of assembly also requires the high cooperativity of non-specific protein interactions, which can be used to achieve accurate control over the co-orientation of protein subunits in nanorings with geometric precision, Figure 3.18b. A C_2-symmetric protein with two district sites, able to chelate divalent metal ions, self-assembles *via* metal coordination and non-covalent interactions in a fixed bending manner into discrete nanorings, the sizes of which can be tuned by varying the ionic strength of the medium.[193] However, a dominating trend in the construction of two-dimensional self-assembly geometries remains the use of wild-type proteins with predefined oligomerization patterns, usually dimerization as it is more straightforward to adapt for higher order structures. The oligomerization of such proteins is controlled not only by metal chelation but also *via* established protein–ligand interactions, which can enable or direct the assembly of oligomerizing proteins into ring-like structures and higher order geometries. An archetypal example is the assembly of homodimer proteins into two-dimensional Pascal triangle lattices.[194] A protein homodimer comprising four domains assembles into triangular ring-shaped structures, which serve as building blocks for the lattices. The formation of the lattices is mediated by interactions between the protein and its native ligand, which ensures the stoichiometric precision of the assembly. Ligand and metal coordination phenomena find extensive use in the construction of cage-like structures. Coordination bonds prove to provide more robust non-covalent binding networks than previously thought. For example, metal coordination bonds at the protein–protein interfaces form cooperatively ordered networks capable of supporting ultrastable artificial protein cages.[195]

Less emphasis in the design of ring-like encapsulating systems is placed on their applications or functional relevance. Protein rings are old forms in Nature's arsenal of functional materials, and are common biological tools, including proteins supporting DNA recombination, *e.g.* RAD52, tryptophan RNA attenuation protein regulating tryptophan synthesis, phage recombinases, DNA helicases and others. The functions that these structures support are highly specialized, and are not easily amenable by design. Conversely, the use of known proteins, which are already assigned to specific functions, limits potential applications. Nevertheless, recent evidence suggests that nanoscale rings can support gene transfer with efficacies comparable to those of closed protein shells. A designed porous cage with a supercharged lumen of ~8 nm was shown to be able to encapsulate siRNA *in vitro*.[196] The cage, comprising six ring-like pores, each of ~3.5 nm, was assembled from 24 highly stable monomers arranged in octahedral geometry.[197] The resulting porous capsule effectively mediated the transfer of its cargo into mammalian cells, releasing it and subsequently inducing the knockdown of model gene expression, Figure 3.18c.

Ultimately, however, it is closed cage-like geometries that offer optimal vehicles for gene delivery.[69] *De novo* assembly topologies using non-viral proteins and artificial polypeptide sequences are being designed to co-assemble with nucleic acids. Coiled-coil sequences were demonstrated to undergo a dendrimer-like assembly, generating remarkably monodisperse hollow shells, which proved to be the smallest virus-like particles among both synthetic and naturally occurring viruses. The particles could readily encapsulate siRNA and DNA, effectively mediating gene silencing and transgene expression,[69] Figure 3.19a. Encapsulation was accompanied by increases in the size of the resulting virions, which is characteristic of native virus shells accommodating different cargo, Figure 3.19a. The important question of whether artificial systems can self-replicate was addressed in an icosahedral shell encapsulating its own mRNA genome.[198] It was possible to evolve virus-like properties for resulting nucleocapsids by generating diversified populations in *Escherichia coli* as an expression host. Several generations of evolution improved genome packaging by more than 133-fold, increased the stability of packaged mRNA in blood from <4% to >70% after 6 h of treatment and prolonged the *in vivo* circulation time from less than 5 min to ~4.5 h, Figure 3.19b.

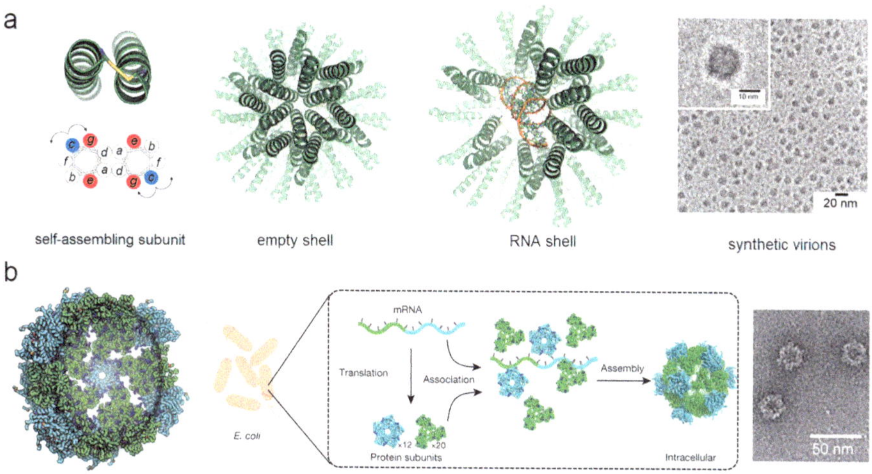

Figure 3.19 Designed RNA virions. (a) A *de novo* virus-like topology assembled from a trifaceted coiled-coil helix. Assemblies are hollow shells, which slightly increase in size when encapsulating RNA. Right: an electron micrograph of empty shells. Reproduced from ref. 69 with permission from American Chemical Society, Copyright 2016. (b) A designed protein shell assembled from trimeric (green) and pentameric (cyan) protein subunits (left), nucleocapsids encapsulating their own mRNA while assembling in *E. coli* cells (middle) and an electron micrograph of the assembly (right). Reproduced from ref. 198 with permission from Springer Nature, Copyright 2017.

The demonstrated ability of such synthetic RNA virions to emulate viruses closely so as to encapsulate functional genes effectively and deliver them into cells and tissues takes us one step closer to therapeutic genome editing.[199] Many other structural designs hold promise for the development of functional nucleocapsids and ultimately self-attenuated artificial viruses expressing tissue-specific genomes. The number of virus-like geometries probed and available for functional evaluations expands to hybrid icosahedral cages revealing high chemical and thermal stabilities,[200] cages assembled from non-cage, membrane-active sequences[201] or virus-derived sequences[202] and markedly accurate icosahedra assembled from precisely defined numbers of protein subunits.[203] In some cases, a successful design may take as little as just one folding motif involved in the assembly of native capsids. For instance, a β-annulus motif common for a variety of viral capsids can be adapted to design a C_3 motif, which spontaneously assembles into discrete nanoscale virus-like particles of sizes comparable to those of their virus precursors, Figure 3.20a and b.[202,204] Other C_3 motifs can derive from combining virus-derived motifs, such as tryptophan zippers,[205] with pharmacophore

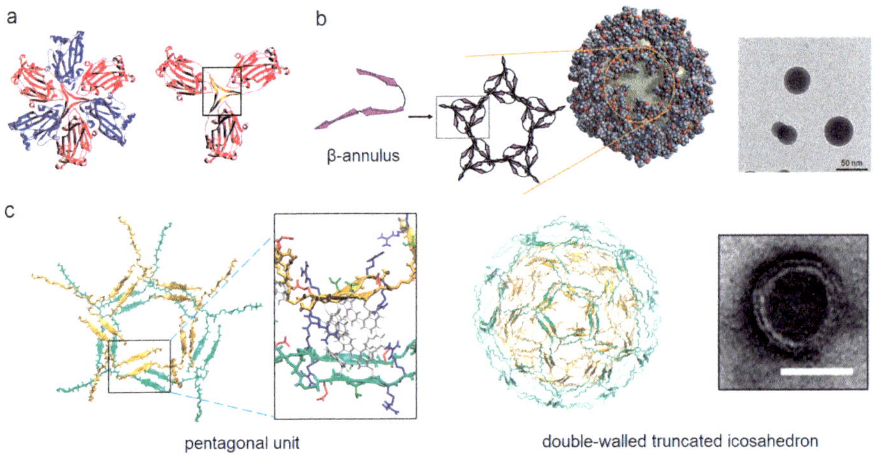

Figure 3.20 Capsid-like assembly of virus-derived C_3 motifs. (a) Hexameric capsomers of Sesbania mosaic virus, with two types of subunits present at the quasi-sixfolds shown in blue and red. A β-annulus structure formed by the N-terminal domains of three subunits of the same type is shown in yellow. Reproduced from ref. 204 with permission from Elsevier, Copyright 2005. (b) Virus-like assembly of a β-annulus motif: schematic model (left) and an electron micrograph (right). Reproduced from ref. 202 with permission from John Wiley & Sons, Copyright © 2010 WILEY-VCH Verlag GmbH & Co. KGaA, Weinheim. (c) A triskel C_3 motif assembling into a pentagonal assembly unit of a double-walled truncated icosahedron: a model (left) and an electron micrograph (right). Outer and inner layers in the icosahedron are shown in green and yellow, respectively. Reproduced from ref. 31 with permission from American Chemical Society, Copyright 2020.

sequences of specialist bacteriostatic proteins, such as lactoferrin.[31] These motifs alone have no appreciable biological activities, but once assembled into capsid-like structures they exhibit strong and rapid antimicrobial responses against resistant and persister bacteria.[31] Intriguingly, the assembled capsids revealed a double-layered arrangement of β-sheet networks stabilized by hydrophobic interfaces running tangentially to the capsid surface, Figure 3.20c, suggesting the formation of peptide bilayers in a manner similar to that of liposomal vesicles, which finds analogies in other designs.[206]

Following a regular cubic symmetry is increasingly taken as a guide for generating discrete folding topologies, which have no analogues in Nature. Such topologies are different from the outlined designs in that they are not hierarchical and tend to fold from a single polypeptide chain. A notable example is a series of coiled-coil designs. These designs are unique in that they utilize complementary coiled-coil segments combined into one contiguous peptide backbone able to fold into individually discrete tetrahedra, both *in vitro* and *in vivo*.[207,208] Strictly, such folds are not encapsulating systems. However, the precision with which they form creates interesting opportunities for designing host–guest systems for ligand or metal coordination, which can be performed in a stoichiometric manner, or as folding reporter systems in living cells.

3.5 Outlook

Overall, nanoscale designs aimed at demonstrating the assembly of particles, vesicles, cages, shells, capsules, capsid, viruses and the like strive to conform to the same principles of host–guest relationships with functional attributes of protection, delivery and compartmentalization. In principle, the pursuit for symmetry-driven designs is relentless. Its very purpose is to diversify cage-like assemblies, with each accommodating a certain number of molecular subunits. The success of a given design is often judged by the accuracy of its structural parameters, *i.e.* how well the target structure is packed, how complementary the designed interfaces are and how conserved and amenable the periphery of the assembled cages is. More elegant and sophisticated structures can be expected to emerge in the near future. However, the functional fitness of encapsulating systems does not strictly rely on structural perfection. Quite the contrary, stable geometries optimized to maintain a desired form may be too rigid and conserved to adapt in dynamic biological environments which determine function. On the other hand, designs approached from a different angle – to find a suitable form for a desired biological function – experience a shift in selection paradigms towards more regular, discrete structures, which symmetry-driven strategies can provide. Therefore, one can expect that designs reminiscent of common, recognizable native forms, such as viruses, but delivering functions different from or unknown for these forms, will become more frequent and better focused on exploitable applications addressing real-life challenges.

References

1. F. Sainsbury, Emergence by Design in Self-Assembling Protein Shells, *ACS Nano*, 2020, **14**(3), 2565–2568.
2. M. M. Conn and J. Rebek, Self-Assembling Capsules, *Chem. Rev.*, 1997, **97**, 1647–1668.
3. F. Hof, S. L. Craig, C. Nuckolls and J. Rebek Jr, Molecular encapsulation, *Angew. Chem., Int. Ed.*, 2002, **41**, 1488–1508.
4. W. M. Stanley and E. G. Valens, *Viruses and Nature of Life*, Dutton, New York, 1971.
5. D. M. Vriezema, M. Comellas Aragonès, J. A. Elemans, J. J. Cornelissen, A. E. Rowan and R. J. Nolte, *Chem. Rev.*, 2005, **105**, 1445–1489.
6. E. G. Percástegui, T. K. Ronson and J. R. Nitschke, Design and Applications of Water-Soluble Coordination Cages, *Chem. Rev.*, 2020, **120**, 13480–13544.
7. E. De Santis and M. G. Ryadnov, Peptide self-assembly for nanomaterials: the old new kid on the block, *Chem. Soc. Rev.*, 2015, **44**, 8288–8300.
8. *Virus Taxonomy:VIIIth Report of the International Committee on the Taxonomy of Viruses*, ed. Fauquet, C. M., Mayo, M. A., Maniloff, J., Desselberger, U. and Ball, L. A., Elsevier Academic Press, 2005.
9. D. Montiel-Garcia, N. Santoyo-Rivera, P. Ho, M. Carrillo-Tripp, J. E. Johnson and V. S. Reddy, VIPERdb v3.0: a structure-based data analytics platform for viral capsids, *Nucleic Acids Res.*, 2021, **49**, D809–D816.
10. F. H. C. Crick and J. D. Watson, Structure of Small Viruses, *Nature*, 1956, **177**, 473–475.
11. J. E. Johnson and J. A. Speir, Quasi-equivalent viruses: a paradigm for protein assemblies, *J. Mol. Biol.*, 1997, **269**, 665–675.
12. R. W. Horne and P. Wildy, Symmetry in virus architecture, *Virology*, 1961, **15**, 348–373.
13. D. L. Caspar and A. Klug, Physical principles in the construction of regular viruses, *Cold Spring Harbor Symp. Quant. Biol.*, 1962, **27**, 1–24.
14. D. L. Caspar and E. Fontano, Five-fold symmetry in crystalline quasicrystal lattices, *Proc. Natl. Acad. Sci. U. S. A.*, 1996, **93**, 14271–14278.
15. X. Yan, *et al.*, Structure and assembly of large lipid-containing dsDNA viruses, *Nat. Struct. Mol. Biol.*, 2000, **7**, 101–103.
16. M. McElwee, S. Vijayakrishnan, F. Rixon and D. Bhella, Structure of the herpes simplex virus portal-vertex, *PLoS Biol.*, 2018, **16**, e2006191.
17. J. T. Finch and A. Klug, Structure of Poliomyelitis Virus, *Nature*, 1959, **183**, 1709–1714.
18. M. G. Rossmann, *et al.*, Structure of a human common cold virus and functional relationship to other picornaviruses, *Nature*, 1985, **317**, 145–153.
19. R. Twarock, Mathematical models for tubular structures in the family of Papovaviridae, *Bull. Math. Biol.*, 2005, **67**, 973–987.

20. R. F. Bruinsma, W. M. Gelbart, D. Reguera, J. Rudnick and R. Zandi, Viral Self-Assembly as a Thermodynamic Process, *Phys. Rev. Lett.*, 2003, **90**, 248101.
21. R. Kerner, A stochastic model of icosahedral capsid growth, *Comput. Math. Methods Med.*, 2005, **6**, 95–97.
22. R. Zandi, P. van der Schoot, D. Reguera, W. Kegel and H. Reiss, Classical Nucleation Theory of Virus Capsids, *Biophys. J.*, 2006, **90**, 1939–1948.
23. R. Twarock, A tiling approach to virus capsid assembly explaining a structural puzzle in virology, *J. Theor. Biol.*, 2004, **226**, 477–482.
24. R. Twarock and A. Luque, Structural puzzles in virology solved with an overarching icosahedral design principle, *Nat. Commun.*, 2019, **10**, 1–9.
25. J. H. Grose, *et al.*, The genomes, proteomes, and structures of three novel phages that infect the bacillus cereus group and carry putative virulence factors, *J. Virol.*, 2014, **88**, 11846–11860.
26. X. Dai and Z. H. Zhou, Structure of the herpes simplex virus 1 capsid with associated tegument protein complexes, *Science*, 2018, **360**, eaao7298.
27. G. C. Lander, *et al.*, Bacteriophage lambda stabilization by auxiliary protein gpD: Timing, location, and mechanism of attachment determined by cryo-EM, *Structure*, 2008, **16**, 1399–1406.
28. E. Mastrobattista, M. A. E. M. van der Aa, W. E. Hennink and D. J. A. Crommelin, Artificial viruses: a nanotechnological approach to gene delivery, *Nat. Rev. Drug Discovery*, 2006, **5**, 115–121.
29. T. O. Yeates, Geometric principles for designing highly symmetric self-assembling protein nanomaterials, *Annu. Rev. Biophys.*, 2017, **46**, 23–42.
30. K. Matsuura, Synthetic approaches to construct viral capsid-like spherical nanomaterials, *Chem. Commun.*, 2018, **54**, 8944–8959.
31. I. E. Kepiro, *et al.*, Engineering Chirally Blind Protein Pseudocapsids into Antibacterial Persisters, *ACS Nano*, 2020, **14**, 1609–1622.
32. H. Feldmann and M. P. Kiley, Classification, structure, and replication of filoviruses, *Curr. Top. Microbiol. Immunol.*, 1999, **235**, 1–21.
33. J. Rakonjac, M. Russel, S. Khanum, S. J. Brooke and M. Rajič, Filamentous Phage: Structure and Biology, *Adv. Exp. Med. Biol.*, 2017, **1053**, 1–20.
34. M. G. Rossmann, Structure of viruses: a short history, *Q. Rev. Biophys.*, 2013, **46**, 133–180.
35. H. Fraenkel-Conrat and R. C. Williams, Reconstitution of active tobacco mosaic virus from its inactive protein and nucleic acid components, *Proc. Natl. Acad. Sci. U. S. A.*, 1955, **41**, 690–698.
36. K. Namba and G. Stubbs, Structure of tobacco mosaic virus at 3.6 A resolution: implications for assembly, *Science*, 1986, **231**, 1401–1406.
37. R. Díaz-Avalos and D. L. D. Caspar, Structure of the Stacked Disk Aggregate of Tobacco Mosaic Virus Protein, *Biophys. J.*, 1998, **74**, 595–603.
38. C. Sachse, *et al.*, High-resolution Electron Microscopy of Helical Specimens: A Fresh Look at Tobacco Mosaic Virus, *J. Mol. Biol.*, 2007, **371**, 812–835.

39. D. J. Kraft, W. K. Kegel and P. van der Schoot, A kinetic zipper model and the assembly of tobacco mosaic virus, *Biophys. J.*, 2012, **102**, 2845–2855.
40. D. L. D. Caspar, Assembly and stability of the tobacco mosaic virus particle, *Adv. Protein Chem.*, 1963, **18**, 37–121.
41. G. P. Martelli, M. J. Adams, J. F. Kreuze and V. V. Dolja, Family Flexiviridae: A Case Study in Virion and Genome Plasticity, *Annu. Rev. Phytopathol.*, 2007, **45**, 73–100.
42. O. Dolnik, L. Kolesnikova and S. Becker, Filoviruses: Interactions with the host cell, *Cell. Mol. Life Sci.*, 2008, **65**, 756–776.
43. C. P. Guyader, B. Lamarre, E. De Santis, J. E. Noble, N. K. Slater and M. G. Ryadnov, Autonomously folded α-helical lockers promote RNAi, *Sci. Rep.*, 2016, **6**, 35012.
44. S. Li, P. Roy, A. Travesset and R. Zandi, Why Large Icosahedral Viruses Need Scaffolding Proteins, *Proc. Natl. Acad. Sci. U. S. A.*, 2018, **115**, 10971–10976.
45. A. Hernandez-Garcia, *et al.*, Design and self-assembly of simple coat proteins for artificial viruses, *Nat. Nanotechnol.*, 2014, **9**, 698–702.
46. D. E. Olins and A. L. Olins, Chromatin history: our view from the bridge, *Nat. Rev. Mol. Cell Biol.*, 2003, **4**, 809–814.
47. K. Luger, A. W. Mader, R. K. Richmond, D. F. Sargent and T. J. Richmond, Crystal structure of the nucleosome core particle at 2.8 A resolution, *Nature*, 1997, **389**, 251–260.
48. S. L. McKnight and O. L. Miller Jr, Ultrastructural patterns of RNA synthesis during early embryogenesis of Drosophila melanogaster, *Cell*, 1976, **8**, 305–319.
49. R. D. Kornberg and J. O. Thomas, Chromatin structure: oligomers of the histones, *Science*, 1974, **184**, 865–868.
50. P. B. Becker and W. Horz, ATP-dependent nucleosome remodeling, *Annu. Rev. Biochem.*, 2002, **71**, 247–273.
51. M. D. Wilson, L. Renault, D. P. Maskell, M. Ghoneim, V. E. Pye, A. Nans, D. S. Rueda, P. Cherepanov and A. Costa, Retroviral integration into nucleosomes through DNA looping and sliding along the histone octamer, *Nat. Commun.*, 2019, **10**, 4189.
52. P. K. Allu, J. M. Dawicki-McKenna, T. Van Eeuwen, M. Slavin, M. Braitbard, C. Xu, N. Kalisman, K. Murakami and B. E. Black, Structure of the Human Core Centromeric Nucleosome Complex, *Curr. Biol.*, 2019, **29**, 2625–2639.
53. G. A. Logsdon, C. W. Gambogi, M. A. Liskovykh, E. J. Barrey, V. Larionov, K. H. Miga, P. Heun and B. E. Black, Human Artificial Chromosomes that Bypass Centromeric DNA, *Cell*, 2019, **178**, 624–639.
54. J. J. Rennick, A. P. R. Johnston and R. G. Parton, Key principles and methods for studying the endocytosis of biological and nanoparticle therapeutics, *Nat. Nanotechnol.*, 2021, **16**, 266–276.
55. T. Kirchhausen, Clathrin, *Annu. Rev. Biochem.*, 2000, **69**, 699–727.
56. K. L. Morris, *et al.*, Cryo-EM of multiple cage architectures reveals a universal mode of clathrin self-assembly, *Nat. Struct. Mol. Biol.*, 2019, **26**, 890–898.

57. M. A. Edeling, C. Smith and D. Owen, Life of a clathrin coat: insights from clathrin and AP structures, *Nat. Rev. Mol. Cell Biol.*, 2006, **7**, 32–44.
58. A. Fotin, *et al.*, Structure of an auxilin-bound clathrin coat and its implications for the mechanism of uncoating, *Nature*, 2004, **432**, 649–653.
59. M. Kaksonen and A. Roux, Mechanisms of clathrin-mediated endocytosis, *Nat. Rev. Mol. Cell Biol.*, 2018, **19**, 313–326.
60. K. Matsuura, K. Murasato and N. Kimizuka, Artificial peptide nanospheres self-assembled from three-way junctions of β-sheet-forming peptides, *J. Am. Chem. Soc.*, 2005, **127**, 10148–10149.
61. M. G. Ryadnov, A Self-Assembling Peptide Polynanoreactor, *Angew. Chem., Int. Ed.*, 2007, **46**, 969–972.
62. S. Ghosh, M. Reches, E. Gazit and S. Verma, Bioinspired Design of Nanocages by Self-Assembling Triskelion Peptide Elements, *Angew. Chem., Int. Ed.*, 2007, **46**, 2002–2004.
63. V. Castelletto, *et al.*, Structurally plastic peptide capsules for synthetic antimicrobial viruses, *Chem. Sci.*, 2016, **7**, 1707–1711.
64. C. M. A. Journot, V. Ramakrishna, M. I. Wallace and A. J. Turberfield, Modifying Membrane Morphology and Interactions with DNA Origami Clathrin-Mimic Networks, *ACS Nano*, 2019, **13**, 9973–9979.
65. E. De Santis, H. Alkassem, B. Lamarre, N. Faruqui, A. Bella, J. E. Noble, N. Micale, S. Ray, J. R. Burns, A. R. Yon, B. W. Hoogenboom and M. G. Ryadnov, Antimicrobial peptide capsids of *de novo* design, *Nat. Commun.*, 2017, **8**, 2263.
66. K. Matsuura, H. Hayashi, K. Murasato and N. Kimizuka, Trigonal tryptophane-zipper as a novel building block for pH-responding peptide nano-assemblies, *Chem. Commun.*, 2011, **47**, 265–267.
67. K. Matsuura, Construction of spherical virus-inspired peptide nanoassemblies, *Polym. J.*, 2012, **44**, 469–474.
68. J. M. Fletcher, *et al.*, Self-assembling cages from coiled-coil peptide modules, *Science*, 2013, **340**, 595–599.
69. J. E. Noble, *et al.*, A *de novo* virus-like topology for synthetic virions, *J. Am. Chem. Soc.*, 2016, **138**, 12202–12210.
70. E. C. Theil, Ferritin: Structure, Gene Regulation, and Cellular Function in Animals, Plants, and Microorganisms, *Annu. Rev. Biochem.*, 1987, **56**, 289–315.
71. H. Huan, Q. Jiang, Y. Wu, X. Qiu, C. Lu, C. Su, J. Zhou, Y. Li, T. Ming and X. Su, Structure determination of ferritin from Dendrorhynchus zhejiangensis, *Biochem. Biophys. Res. Commun.*, 2020, **531**, 195–202.
72. P. M. Harrison and P. Arosio, The ferritins: molecular properties, iron storage function and cellular regulation, *Biochim. Biophys. Acta*, 1996, **1275**, 161–203.
73. O. Kasyutich, A. Ilari, A. Fiorillo, D. Tatchev, A. Hoell and P. Ceci, Silver ion incorporation and nanoparticle formation inside the cavity of Pyrococcus furiosus ferritin: structural and size-distribution analyses, *J. Am. Chem. Soc.*, 2010, **132**, 3621–3627.
74. G. Jutz, P. van Rijn, B. Santos Miranda and A. Böker, Ferritin: a versatile building block for bionanotechnology, *Chem. Rev.*, 2015, **115**, 1653–1701.

75. S. Mann, *Biomineralization: Principles and Concepts in Bioinorganic Materials Chemistry*, Oxford University Press, Oxford, 2002.
76. A. Treffry and P. M. Harrison, Spectroscopic studies on the binding of Iron, Terbium, and Zinc by Apoferritin, *J. Inorg. Biochem.*, 1984, **21**, 9–20.
77. C. M. Barnes, E. C. Theil and K. N. Raymond, Iron uptake in ferritin is blocked by binding of $[Cr(TREN)(H2O)(OH)]^{2+}$, a slow dissociating model for $[Fe(H2O)6]^{2+}$, *Proc. Natl. Acad. Sci. U. S. A.*, 2002, **99**, 5195–5200.
78. X. Yang, P. Arosio and N. D. Chasteen, Molecular Diffusion into Ferritin: Pathways, Temperature Dependence, Incubation Time, and Concentration Effects, *Biophys. J.*, 2000, **78**, 2049–2059.
79. F. C. Meldrum, V. J. Wade, D. L. Nimmo, B. R. Heywood and S. Mann, Synthesis of inorganic nanophase materials in supramolecular protein cages, *Nature*, 1991, **349**, 684–687.
80. T. Douglas, *et al.*, Synthesis and Structure of an Iron(III) Sulfide-Ferritin Bioinorganic Nanocomposite, *Science*, 1995, **269**, 54–57.
81. S. Gider, D. Awschalom, T. Douglas, S. Mann and M. Chaparala, Classical and quantum magnetic phenomena in natural and artificial ferritin proteins, *Science*, 1995, **268**, 77–80.
82. F. C. Meldrum and S. Mann, Controlled synthesis of inorganic materials using supramolecular assemblies, *Adv. Mater.*, 1991, **30**, 316–318.
83. F. Meldrum, B. Heywood and S. Mann, Magnetoferritin: *in vitro* synthesis of a novel magnetic protein, *Science*, 1992, **257**, 522–523.
84. R. M. Kramer, C. Li, D. C. Carter, M. O. Stone and R. R. Naik, Engineered Protein Cages for Nanomaterial Synthesis, *J. Am. Chem. Soc.*, 2004, **126**, 13282–13286.
85. R. R. Naik, S. J. Stringer, G. Agarwal, S. E. Jones and M. O. Stone, Biomimetic synthesis and patterning of silver nanoparticles, *Nat. Mater.*, 2002, **1**, 169–172.
86. S. R. Whaley, D. S. English, E. L. Hu, P. F. Barbara and A. M. Belcher, Selection of peptides with semiconductor binding specificity for directed nanocrystal assembly, *Nature*, 2000, **405**, 665–668.
87. I. Moglia, M. Santiago, M. Soler and A. Olivera-Nappa, Silver nanoparticle synthesis in human ferritin by photochemical reduction, *J. Inorg. Biochem.*, 2020, **206**, 111016.
88. D. Xu, G. D. Watt, J. N. Harb and R. C. Davis, Electrical Conductivity of Ferritin Proteins by Conductive AFM, *Nano Lett.*, 2005, **5**, 571–577.
89. I. Yamashita, K. Iwahori and S. Kumagai, Ferritin in the field of nanodevices, *Biochim. Biophys. Acta*, 2010, **1800**, 846–857.
90. V. Nandwana, *et al.*, Engineered ferritin nanocages as natural contrast agents in magnetic resonance imaging, *RSC Adv.*, 2017, **7**, 34892–34900.
91. C. Batya, D. Hagit, M. Gila, H. Alon and N. Michal, Ferritin as an Endogenous MRI Reporter for Noninvasive Imaging of Gene Expression in C6 Glioma Tumors, *Neoplasia*, 2005, **7**, 109–117.

92. T. Douglas and A. Young, Viruses: making friends with old foes, *Science*, 2006, **312**, 873–875.
93. J. A. Speir, S. Munshi, G. Wang, T. S. Baker and J. E. Johnson, Structures of the native and swollen forms of cowpea chlorotic mottle virus determined by X-ray crystallography and cryo-electron microscopy, *Structure*, 1995, **3**, 63–78.
94. T. Lin, *et al.*, The Refined Crystal Structure of Cowpea Mosaic Virus at 2.8 Å Resolution, *Virology*, 1999, **265**, 20–34.
95. B. D. Wilts, I. A. T. Schaap and C. F. Schmidt, Swelling and softening of the cowpea chlorotic mottle virus in response to pH shifts, *Biophys. J.*, 2015, **108**, 2541–2549.
96. T. Douglas and M. Young, Host-guest encapsulation of materials by assembled virus protein cages, *Nature*, 1998, **393**, 152–155.
97. T. Douglas, *et al.*, Protein engineering of a viral cage for constrained nanomaterials synthesis, *Adv. Mater.*, 2002, **14**, 415–418.
98. M. T. Klem, D. Willits, M. Young and T. Douglas, 2-D Array formation of genetically engineered viral cages on Au surfaces and imaging by atomic force microscopy, *J. Am. Chem. Soc.*, 2003, **125**, 10806–10807.
99. J. N. Culver, A. D. Brown, F. Zang, M. Gnerlich, K. Gerasopoulos and R. Ghodssi, Plant virus directed fabrication of nanoscale materials and devices, *Virology*, 2015, **479–480**, 200–212.
100. M. Uchida, *et al.*, Biological containers: protein cages as multifunctional nanoplatforms, *Adv. Mater.*, 2007, **19**, 1025–1042.
101. Q. Wang, E. Kaltgrad, T. Lin, J. E. Johnson and M. G. Finn, Natural Supramolecular Building Blocks: Wild-Type Cowpea Mosaic Virus, *Chem. Biol.*, 2002, **9**, 805–811.
102. A. Chatterji, *et al.*, New Addresses on an Addressable Virus Nanoblock: Uniquely Reactive Lys Residues on Cowpea Mosaic Virus, *Chem. Biol.*, 2004, **11**, 855–863.
103. C. M. Soto, *et al.*, Fluorescent signal amplification of carbocyanine dyes using engineered viral nanoparticles, *J. Am. Chem. Soc.*, 2006, **128**, 5184–5189.
104. P. H. Beatty and J. D. Lewis, Cowpea mosaic virus nanoparticles for cancer imaging and therapy, *Adv. Drug Delivery Rev.*, 2019, **145**, 130–144.
105. S. E. Vieweger, I. B. Tsvetkova and B. G. Dragnea, *In Vitro* Assembly of Virus-Derived Designer Shells Around Inorganic Nanoparticles, *Methods Mol. Biol.*, 2018, **1776**, 279–294.
106. C. Zeng, G. Rodriguez Lázaro, I. B. Tsvetkova, M. F. Hagan and B. Dragnea, Defects and Chirality in the Nanoparticle-Directed Assembly of Spherocylindrical Shells of Virus Coat Proteins, *ACS Nano*, 2018, **12**, 5323–5332.
107. M. V. de Ruiter, R. M. van der Hee, A. J. M. Driessen, E. D. Keurhorst, M. Hamid and J. J. L. M. Cornelissen, Polymorphic assembly of virus-capsid proteins around DNA and the cellular uptake of the resulting particles, *J. Controlled Release*, 2019, **307**, 342–354.

108. S. Shukla, C. Wang, V. Beiss, H. Cai, T. Washington 2nd, A. A. Murray, X. Gong, Z. Zhao, H. Masarapu, A. Zlotnick, S. Fiering and N. F. Steinmetz, The unique potency of Cowpea mosaic virus (CPMV) *in situ* cancer vaccine, *Biomater. Sci.*, 2020, **8**, 5489–5503.
109. P. H. Lizotte, A. M. Wen, M. R. Sheen, J. Fields, P. Rojanasopondist, N. F. Steinmetz and S. Fiering, In situ vaccination with cowpea mosaic virus nanoparticles suppresses metastatic cancer, *Nat. Nanotechnol.*, 2016, **11**, 295–303.
110. J. D. Lewis, *et al.*, Viral nanoparticles as tools for intravital vascular imaging, *Nat. Med.*, 2006, **12**, 354–360.
111. E. Carlee, *et al.*, Cell-Specific Delivery of Diverse Cargos by Bacteriophage MS2 Virus-like Particles, *ACS Nano*, 2011, **5**, 5729–5745.
112. H.-A. Nguyen, *et al.*, Assembly of gold nanoparticles using turnip yellow mosaic virus as an in-solution SERS sensor, *RSC Adv.*, 2019, **9**, 32296–32307.
113. H. Yi, *et al.*, Patterned assembly of genetically modified viral nanotemplates *via* nucleic acid hybridization, *Nano Lett.*, 2005, **5**, 1931–1936.
114. T. L. Schlick, Z. Ding, E. W. Kovacs and M. B. Francis, Dual-Surface Modification of the Tobacco Mosaic Virus, *J. Am. Chem. Soc.*, 2005, **127**, 3718–3723.
115. E. Dujardin, C. Peet, G. Stubbs, J. N. Culver and S. Mann, Organization of Metallic Nanoparticles Using Tobacco Mosaic Virus Templates, *Nano Lett.*, 2003, **3**, 413–417.
116. M. Knez, *et al.*, Biotemplate Synthesis of 3-nm Nickel and Cobalt Nanowires, *Nano Lett.*, 2003, **3**, 1079–1082.
117. T. R. Walsh and M. R. Knecht, Biointerface Structural Effects on the Properties and Applications of Bioinspired Peptide-Based Nanomaterials, *Chem. Rev.*, 2017, **117**, 12641–12704.
118. M. Kobayashi, M. Seki, H. Tabata, Y. Watanabe and I. Yamashita, Fabrication of aligned magnetic nanoparticles using tobamoviruses, *Nano Lett.*, 2010, **10**, 773–776.
119. A. Nedoluzhko and T. Douglas, Ordered association of tobacco mosaic virus in the presence of divalent metal ions, *J. Inorg. Biochem.*, 2001, **84**, 233–240.
120. S.-W. Lee, C. Mao, C. E. Flynn and A. M. Belcher, Ordering of Quantum Dots Using Genetically Engineered Viruses, *Science*, 2002, **296**, 892–895.
121. K. T. Nam, *et al.*, Virus-Enabled Synthesis and Assembly of Nanowires for Lithium Ion Battery Electrodes, *Science*, 2006, **312**, 885–888.
122. L. Yang, *et al.*, Self-assembled virus-membrane complexes, *Nat. Mater.*, 2004, **3**, 615–619.
123. W. J. Chung, J. W. Oh, K. Kwak, B. Y. Lee, J. Meyer, E. Wang, A. Hexemer and S. W. Lee, Biomimetic self-templating supramolecular structures, *Nature*, 2011, **478**, 364–368.

124. S. Y. Yoo, H. E. Jin, D. S. Choi, M. Kobayashi, Y. Farouz, S. Wang and S. W. Lee, M13 Bacteriophage and Adeno-Associated Virus Hybrid for Novel Tissue Engineering Matcrial with Gene Delivery Functions, *Adv. Healthcare Mater.*, 2016, **5**, 88–93.
125. J. H. Lee, C. M. Warner, H. E. Jin, E. Barnes, A. R. Poda, E. J. Perkins and S. W. Lee, Production of tunable nanomaterials using hierarchically assembled bacteriophages, *Nat. Protoc.*, 2017, **12**, 1999–2013.
126. J. W. Oh, W. J. Chung, K. Heo, H. E. Jin, B. Y. Lee, E. Wang, C. Zueger, W. Wong, J. Meyer, C. Kim, S. Y. Lee, W. G. Kim, M. Zemla, M. Auer, A. Hexemer and S. W. Lee, Biomimetic virus-based colourimetric sensors, *Nat. Commun.*, 2014, **5**, 3043.
127. C. E. Dunbar, K. A. High, J. K. Joung, D. B. Kohn, K. Ozawa and M. Sadelain, Gene therapy comes of age, *Science*, 2018, **359**, eaan4672.
128. S. L. Ginn, A. K. Amaya, I. E. Alexander, M. Edelstein and M. R. Abedi, Gene therapy clinical trials worldwide to 2017: An update, *J. Gene Med.*, 2018, **20**, e3015.
129. N. A. Alhakamy, D. T. Curiel and C. J. Berkland, The era of gene therapy: from preclinical development to clinical application, *Drug Discovery Today*, 2021, **26**, 1602–1619.
130. E. A. Hurley, D. Hull and S. P. Shriver, The Next Phase of Human Gene-Therapy Oversight, *N. Engl. J. Med.*, 2019, **380**, 401–402.
131. M. W. Nirenberg, Will Society be prepared?, *Science*, 1967, **157**, 633.
132. H. Yin, R. L. Kanasty, A. A. Eltoukhy, A. J. Vegas, J. R. Dorkin and D. G. Anderson, Non-viral vectors for gene-based therapy, *Nat. Rev. Genet.*, 2014, **15**, 541–555.
133. M. A. Mintzer and E. E. Simanek, Nonviral vectors for gene delivery, *Chem. Rev.*, 2009, **109**, 259–302.
134. S. Mehier-Humbert and R. H. Guy, Physical methods for gene transfer: Improving the kinetics of gene delivery into cells, *Adv. Drug Delivery Rev.*, 2005, **57**, 733–753.
135. J. L. Young and D. A. Dean, Electroporation-mediated gene delivery, *Adv. Genet.*, 2015, **89**, 49–88.
136. J.-I. Miyazaki and H. Aihara, *Gene Ther. Protoc.*, 2002, 49–62.
137. R. Waehler, S. J. Russell and D. T. Curiel, Engineering targeted viral vectors for gene therapy, *Nat. Rev. Genet.*, 2007, **8**, 573–587.
138. D. J. Glover, H. J. Lipps and D. A. Jans, Towards safe, non-viral therapeutic gene expression in humans, *Nat. Rev. Genet.*, 2005, **6**, 299–310.
139. B. Lamarre and M. G. Ryadnov, Self-assembling viral mimetics: one long journey with short steps, *Macromol. Biosci.*, 2011, **11**, 503–513.
140. I. Kopatz, *et al.*, A model for non-viral gene delivery: through syndecan adhesion molecules and powered by actin, *J. Gene Med.*, 2004, **6**, 769–776.
141. D. Oupicky, A. L. Parker and L. W. Seymour, Laterally Stabilized Complexes of DNA with Linear Reducible Polycations: Strategy for Triggered Intracellular Activation of DNA Delivery Vectors, *J. Am. Chem. Soc.*, 2002, **124**, 8–9.

142. P. Kos, U. Lächelt, A. Herrmann, F. M. Mickler, M. Döblinger, D. He, A. Krhač Levačić, S. Morys, C. Bräuchle and E. Wagner, Histidine-rich stabilized polyplexes for cMet-directed tumor-targeted gene transfer, *Nanoscale*, 2015, **7**, 5350–5362.
143. Z. Kadlecova, L. Baldi, D. Hacker, F. M. Wurm and H.-A. Klok, Comparative study on the *in vitro* cytotoxicity of linear, dendritic, and hyperbranched polylysine analogues, *Biomacromolecules*, 2012, **13**, 3127–3137.
144. S. T. Crowley, J. A. Poliskey, N. J. Baumhover and K. G. Rice, Efficient expression of stabilized mRNA PEG-peptide polyplexes in liver, *Gene Ther.*, 2015, **22**, 993–999.
145. S. Uchida, K. Itaka, Q. Chen, K. Osada, T. Ishii, M. A. Shibata, M. Harada-Shiba and K. Kataoka, PEGylated polyplex with optimized PEG shielding enhances gene introduction in lungs by minimizing inflammatory responses, *Mol. Ther.*, 2012, **20**, 1196–1203.
146. A. Kolate, D. Baradia, S. Patil, I. Vhora, G. Kore and A. Misra, PEG - a versatile conjugating ligand for drugs and drug delivery systems, *J. Controlled Release*, 2014, **192**, 67–81.
147. E. Bertosin, P. Stömmer, E. Feigl, M. Wenig, M. N. Honemann and H. Dietz, Cryo-Electron Microscopy and Mass Analysis of Oligolysine-Coated DNA Nanostructures, *ACS Nano*, 2021, 9391–9403.
148. M. Suzuki, SPKK, a new nucleic acid-binding unit of protein found in histone, *EMBO J.*, 1989, **8**, 797–804.
149. H. Han, J. Yang, W. Chen, Q. Li, Y. Yang and Q. Li, A comprehensive review on histone-mediated transfection for gene therapy, *Biotechnol. Adv.*, 2019, **37**, 132–144.
150. M. M. Munye, J. Ravi, A. D. Tagalakis, D. McCarthy, M. G. Ryadnov and S. L. Hart, Role of liposome and peptide in the synergistic enhancement of transfection with a lipopolyplex vector, *Sci. Rep.*, 2015, **5**, 9292.
151. J. Cheng, R. Zeidan, S. Mishra, A. Liu, S. H. Pun, R. P. Kulkarni, G. S. Jensen, N. C. Bellocq and M. E. Davis, Structure-function correlation of chloroquine and analogues as transgene expression enhancers in non-viral gene delivery, *J. Med. Chem.*, 2006, **49**, 6522–6531.
152. M. Gestin, M. Dowaidar and Ü. Langel, Uptake Mechanism of Cell-Penetrating Peptides, *Adv. Exp. Med. Biol.*, 2017, **1030**, 255–264.
153. D. Derossi, A. H. Joliot, G. Chassaing and A. Prochiantz, The third helix of the Antennapedia homeodomain translocates through biological membranes, *J. Biol. Chem.*, 1994, **269**, 10444–10450.
154. E. Vives, P. Brodin and B. Lebleu, A Truncated HIV-1 Tat Protein Basic Domain Rapidly Translocates through the Plasma Membrane and Accumulates in the Cell Nucleus, *J. Biol. Chem.*, 1997, **272**, 16010–16017.
155. G. Elliott and P. O'Hare, Intercellular Trafficking and Protein Delivery by a Herpesvirus Structural Protein, *Cell*, 1997, **88**, 223–233.
156. S. Boeckle, J. Fahrmeir, W. Roedl, M. Ogris and E. Wagner, Melittin analogs with high lytic activity at endosomal pH enhance transfection with purified targeted PEI polyplexes, *J. Controlled Release*, 2006, **112**, 240–248.

157. E. Wagner, C. Plank, K. Zatloukal, M. Cotten and M. L. Birnstiel, Influenza virus hemagglutinin HA-2 N-terminal fusogenic peptides augment gene transfer by transferrin-polylysine-DNA complexes: toward a synthetic virus-like gene-transfer vehicle, *Proc. Natl. Acad. Sci. U. S. A.*, 1992, **89**, 7934–7938.
158. S. Gottschalk, *et al.*, A novel DNA-peptide complex for efficient gene transfer and expression in mammalian cells, *Gene Ther.*, 1996, **3**, 448–457.
159. S. M. Hammond, G. Hazell, F. Shabanpoor, A. F. Saleh, M. Bowerman, J. N. Sleigh, K. E. Meijboom, H. Zhou, F. Muntoni, K. Talbot, M. J. Gait and M. J. Wood, Systemic peptide-mediated oligonucleotide therapy improves long-term survival in spinal muscular atrophy, *Proc. Natl. Acad. Sci. U. S. A.*, 2016, **113**, 10962–10967.
160. M. D. Pierschbacher and E. Ruoslahti, Cell attachment activity of fibronectin can be duplicated by small synthetic fragments of the molecule, *Nature*, 1984, **309**, 30–33.
161. J. Graf, *et al.*, Identification of an amino acid sequence in laminin mediating cell attachment, chemotaxis, and receptor binding, *Cell*, 1987, **48**, 989–996.
162. T. W. R. Lee, G. E. Blair and D. A. Matthews, Adenovirus core protein VII contains distinct sequences that mediate targeting to the nucleus and nucleolus, and colocalization with human chromosomes, *J. Gen. Virol.*, 2003, **84**, 3423–3428.
163. A. D. Bangham, M. M. Standish and J. C. Watkins, Diffusion of univalent ions across the lamellae of swollen phospholipids, *J. Mol. Biol.*, 1965, **13**, 238–252.
164. P. L. Felgner, *et al.*, Lipofection: a highly efficient, lipid-mediated DNA-transfection procedure, *Proc. Natl. Acad. Sci. U. S. A.*, 1987, **84**, 7413–7417.
165. N. S. Templeton, *et al.*, Improved DNA: liposome complexes for increased systemic delivery and gene expression, *Nat. Biotechnol.*, 1997, **15**, 647–652.
166. I. Koltover, T. Salditt, J. O. Radler and C. R. Safinya, An Inverted Hexagonal Phase of Cationic Liposome-DNA Complexes Related to DNA Release and Delivery, *Science*, 1998, **281**, 78–81.
167. C. R. Safinya, Structures of lipid-DNA complexes: supramolecular assembly and gene delivery, *Curr. Opin. Struct. Biol.*, 2001, **11**, 440–448.
168. H.-J. Butt, K. Graf and M. Kappl, *Physics and Chemistry of Interfaces*, Wiley-VCH, Weinheim, 2003.
169. E. Wonder, L. Simón-Gracia, P. Scodeller, R. N. Majzoub, V. R. Kotamraju, K. K. Ewert, T. Teesalu and C. R. Safinya, Competition of charge-mediated and specific binding by peptide-tagged cationic liposome-DNA nanoparticles *in vitro* and *in vivo*, *Biomaterials*, 2018, **166**, 52–63.
170. T. M. Reineke, Stimuli-Responsive Polymers for Biological Detection and Delivery, *ACS Macro Lett.*, 2016, **5**, 14–18.
171. M. P. Stewart, R. Langer and K. F. Jensen, Intracellular Delivery by Membrane Disruption: Mechanisms, Strategies, and Concepts, *Chem. Rev.*, 2018, **118**, 7409–7531.

172. I. Lostale-Seijo and J. Montenegro, Synthetic materials at the forefront of gene delivery, *Nat. Rev. Chem.*, 2018, **2**, 258–277.
173. M. X. Tang, W. Li and F. C. Szoka, Toroid formation in charge neutralized flexible or semi-flexible biopolymers: potential pathway for assembly of DNA carriers, *J. Gene Med.*, 2005, **7**, 334–342.
174. J. O. Radler, I. Koltover, T. Salditt and C. R. Safinya, Structure of DNA-Cationic Liposome Complexes: DNA Intercalation in Multilamellar Membranes in Distinct Interhelical Packing Regimes, *Science*, 1997, **275**, 810–814.
175. C. Chittimalla, L. Zammut-Italiano, G. Zuber and J. P. Behr, Monomolecular DNA Nanoparticles for Intravenous Delivery of Genes, *J. Am. Chem. Soc.*, 2005, **127**, 11436–11441.
176. Y. Aoyama, *et al.*, Artificial Viruses and Their Application to Gene Delivery. Size-Controlled Gene Coating with Glycocluster Nanoparticles, *J. Am. Chem. Soc.*, 2003, **125**, 3455–3457.
177. T. Nakai, T. Kanamori, S. Sando and Y. Aoyama, Remarkably Size-Regulated Cell Invasion by Artificial Viruses. Saccharide-Dependent Self-Aggregation of Glycoviruses and Its Consequences in Glycoviral Gene Delivery, *J. Am. Chem. Soc.*, 2003, **125**, 8465–8475.
178. B. T. Michen and T. Graule, Isoelectric points of viruses, *J. Appl. Microbiol.*, 2010, **109**, 388–397.
179. E. P. Holowka, V. Z. Sun, D. T. Kamei and T. J. Deming, Polyarginine segments in block copolypeptides drive both vesicular assembly and intracellular delivery, *Nat. Mater.*, 2007, **6**, 52–57.
180. E. K. Chung, E. Lee, Y. B. Lim and M. Lee, Cyclic peptide facial amphiphile preprogrammed to self-assemble into bioactive peptide capsules, *Chem. - Eur. J.*, 2010, **16**, 5305–5309.
181. J. E. Padilla, C. Colovos and T. O. Yeates, Nanohedra: Using symmetry to design self-assembling protein cages, layers, crystals, and filaments, *Proc. Natl. Acad. Sci. U. S. A.*, 2001, **98**, 2217–2221.
182. J. Laniado, K. A. Cannon, J. E. Miller, M. R. Sawaya, D. E. McNamara and T. O. Yeates, Geometric Lessons and Design Strategies for Nanoscale Protein Cages, *ACS Nano*, 2021, **15**, 4277–4286.
183. S. Raman, G. Machaidze, A. Lustig, U. Aebi and P. Burkhard, Structure-based design of peptides that self-assemble into regular polyhedral nanoparticles, *Nanomedicine*, 2006, **2**, 95–102.
184. C. P. Karch, H. Bai, O. B. Torres, C. A. Tucker, N. L. Michael, G. R. Matyas, M. Rolland, P. Burkhard and Z. Beck, Design and characterization of a self-assembling protein nanoparticle displaying HIV-1 Env V1V2 loop in a native-like trimeric conformation as vaccine antigen, *Nanomedicine*, 2019, **16**, 206–216.
185. M. G. Ryadnov, A self-assembling peptide polynanoreactor, *Angew. Chem., Int. Ed.*, 2007, **46**, 969–972.
186. S. C. Zimmerman, F. Zeng, D. E. C. Reichert and S. V. Kolotuchin, Self-Assembling Dendrimers, *Science*, 1996, **271**, 1095–1098.

187. D. A. Tomalia, H. M. Brothers, L. T. Piehler, H. D. Durst and D. R. Swanson, Partial shell-filled core-shell tecto(dendrimers): A strategy to surface differentiated nano-clefts and cusps, *Proc. Natl. Acad. Sci. U. S. A.*, 2002, **99**, 5081–5087.
188. S. Hecht and J. M. J. Fréchet, Dendritic Encapsulation of Function: Applying Nature's Site Isolation Principle from Biomimetics to Materials Science, *Angew. Chem., Int. Ed.*, 2001, **40**, 74–91.
189. Z. Lyu, L. Ding, A. Tintaru and L. Peng, Self-Assembling Supramolecular Dendrimers for Biomedical Applications: Lessons Learned from Poly(amidoamine) Dendrimers, *Acc. Chem. Res.*, 2020, **53**, 2936–2949.
190. P. S. Kowalski, A. Rudra, L. Miao and D. G. Anderson, Delivering the Messenger: Advances in Technologies for Therapeutic mRNA Delivery, *Mol. Ther.*, 2019, **27**, 710–728.
191. C. Rodriguez-Emmenegger, *et al.*, Encoding biological recognition in a bicomponent cell-membrane mimic, *Proc. Natl. Acad. Sci. U. S. A.*, 2019, **116**, 5376–5382.
192. J. C. T. Carlson, *et al.*, Chemically Controlled Self-Assembly of Protein Nanorings, *J. Am. Chem. Soc.*, 2006, **128**, 7630–7638.
193. Y. Bai, Q. Luo, W. Zhang, L. Miao, J. Xu, H. Li and J. Liu, Highly ordered protein nanorings designed by accurate control of glutathione S-transferase self-assembly, *J. Am. Chem. Soc.*, 2013, **135**, 10966–10969.
194. R. Liu, Z. Kochovski, L. Li, Y. W. Yin, J. Yang, G. Yang, G. Tao, A. Xu, E. Zhang, H. M. Ding, Y. Lu, G. Chen and M. Jiang, Fabrication of Pascal-triangle Lattice of Proteins by Inducing Ligand Strategy, *Angew. Chem., Int. Ed.*, 2020, **59**, 9617–9623.
195. A. D. Malay, *et al.*, An ultra-stable gold-coordinated protein cage displaying reversible assembly, *Nature*, 2019, **569**, 438–442.
196. T. G. W. Edwardson, T. Mori and D. Hilvert, Rational Engineering of a Designed Protein Cage for siRNA Delivery, *J. Am. Chem. Soc.*, 2018, **140**, 10439–10442.
197. N. P. King, W. Sheffler, M. R. Sawaya, B. S. Vollmar, J. P. Sumida, I. André, T. Gonen, T. O. Yeates and D. Baker, Computational design of self-assembling protein nanomaterials with atomic level accuracy, *Science*, 2012, **336**, 1171.
198. G. L. Butterfield, *et al.*, Evolution of a designed protein assembly encapsulating its own RNA genome, *Nature*, 2017, **552**, 415–420.
199. J. van Haasteren, J. Li, O. J. Scheideler, N. Murthy and D. V. Schaffer, The delivery challenge: fulfilling the promise of therapeutic genome editing, *Nat. Biotechnol.*, 2020, **38**, 845–855.
200. A. S. Cristie-David, *et al.*, Coiled-coil-mediated assembly of an icosahedral protein cage with extremely high thermal and chemical stability, *J. Am. Chem. Soc.*, 2019, **141**, 9207–9216.
201. S. G. Tarasov, V. Gaponenko, O. M. Howard, Y. Chen, J. J. Oppenheim, M. A. Dyba, S. Subramaniam, Y. Lee, C. Michejda and N. I. Tarasova, Structural plasticity of a transmembrane peptide allows self-assembly

into biologically active nanoparticles, *Proc. Natl. Acad. Sci. U. S. A.*, 2011, **108**, 9798–9803.
202. K. Matsuura, K. Watanabe, T. Matsuzaki, K. Sakurai and N. Kimizuka, Self-assembled synthetic viral capsids from a 24-mer viral peptide fragment, *Angew. Chem., Int. Ed.*, 2010, **49**, 9662–9665.
203. J. B. Bale, *et al.*, Accurate design of megadalton-scale two-component icosahedral protein complexes, *Science*, 2016, **353**, 389–394.
204. P. S. Satheshkumar, G. L. Lokesh, M. R. Murthy and H. S. Savithri, The role of arginine-rich motif and beta-annulus in the assembly and stability of Sesbania mosaic virus capsids, *J. Mol. Biol.*, 2005, **353**, 447–458.
205. A. G. Cochran, N. J. Skelton and M. A. Starovasnik, Tryptophan zippers: stable, monomeric beta -hairpins, *Proc. Natl. Acad. Sci. U. S. A.*, 2001, **98**, 5578–5583.
206. M. Rad-Malekshahi, K. M. Visscher, J. P. Rodrigues, R. De Vries, W. E. Hennink, M. Baldus, A. M. Bonvin, E. Mastrobattista and M. Weingarth, The supramolecular organization of a peptide-based nanocarrier at high molecular detail, *J. Am. Chem. Soc.*, 2015, **137**, 7775–7784.
207. A. Ljubetič, *et al.*, Design of coiled-coil protein-origami cages that self-assemble *in vitro* and *in vivo*, *Nat. Biotechnol.*, 2017, **35**, 1094–1101.
208. H. Gradišar, *et al.*, Design of a single-chain polypeptide tetrahedron assembled from coiled-coil segments, *Nat. Chem. Biol.*, 2013, **9**, 362–366.

CHAPTER 4

Nanoscale Designs for Tissue Regeneration

4.1 Introduction

Regeneration has been an everlasting dream of human society. There is hardly another area of interest and importance that would cause excitement while feeding scepticism.[1,2] However remote it may seem from clinic, regeneration is ubiquitous to all multicellular organisms. Questions remain as to why some forms of life have such an ability to restore lost or damaged tissue or body parts and others do not. Regeneration may appeal to the rational mind as an obvious solution to many biomedical problems. However, it is rejected for humans as evolutionarily disadvantageous.[3,4] But is it different to other forms of life? It is not that common for plants to regenerate fully from individual cells. Only a small number of invertebrates can be reborn from body parts, and not all, but some, amphibians can restore lost limbs. Regeneration appears to be tuned to a degree unique to each species. This gives hope and inspires us to attempt to put the mechanisms of this phenomenon to medical use.[5]

Regeneration maintains the integrity of tissues and organs, which is consistent and equally vital for all organisms, including those with seemingly lower regenerative capabilities – us. Typical examples are skin and blood that undergo complete cellular renewal an innumerable number of times during our life span. This type of regeneration is maintenance regeneration.[6]

In this process, reserves of stem cells maintain constant contact with their environment to be ready to respond to cues activating their differentiation into a specific tissue. In contrast, tissue damage is dealt with by injury-induced regeneration. Injured tissues require much faster regeneration

Nanoscience & Nanotechnology Series No. 53
Bionanodesign: Old Forms for New Functions, 2nd Edition
By Maxim Ryadnov

Published by the Royal Society of Chemistry, www.rsc.org

turnover, mobilizing stem cells *via* partial reversal of their original embryonic development, which involves regression of specialized cells, *i.e.* cells already differentiated into a tissue, into unspecialized cells, *i.e.* stem cells. This process is termed dedifferentiation, involving the formation of an unspecialized tissue at an injury site to serve as a regrowth bud.[7]

Dedifferentiation shows remarkable efficiency in plants and invertebrates in repairing damaged or missing tissues and organs. Low vertebrates such as urodele amphibians can renew lost or severed body parts relying on similar mechanisms.[8] Mammalian cells do not regenerate *via* dedifferentiation, an ability believed to have long been lost as a result of evolution. However, wound repair and tissue regeneration are inherent to mammals, including humans. Healing is provided by another mechanism – fibrosis.[9]

Fibrosis occurs as a result of an inflammatory response accompanied by the formation of large amounts of fibrous tissues that undergo remodelling and maturation into scars. It can also be triggered in healing extensive wounds, the size of which is beyond the regenerative capacity of regeneration-competent tissues. Strictly speaking, fibrosis is not a regeneration mechanism as it does not restore native tissues. Moreover, given that many tissues viewed as incapable of regeneration do contain regeneration-competent stem cells, scarring appears as a consequence of suppressed regeneration.[10] Although exact relationships between regeneration and fibrosis remain to be established, both of these phenomena underlie one of the main strategies for advancing regenerative medicine – the development of synthetic analogues of the extracellular matrix (ECM). The strategy stems from the excessive production of ECM during tissue repair and fibrosis in mammals, Figure 4.1.[11] With the ECM being a prerequisite for the proliferation of most cells, synthetic ECM analogues represent an attractive aim for nanoscale biodesigns.[12]

4.2 The Extracellular Matrix (ECM) – Its Role and Composition

Multicellular organisms produce extracellular matrices as tissue scaffolds, Figure 4.2a.[13] The native ECM is a collagenous mesh embedding matrix glycoproteins, growth factors, metalloproteinases and other macromolecular mediators of matrix–cell interactions.

The role of the ECM in tissue development is twofold. First, the ECM provides tensile strength to a tissue, and defines its 3D architecture through the formation of complex supramolecular structures. Second, such structures promote cell adhesion and migration, provide a storage depot of growth factors and enable signalling for cellular homeostasis.[14] The ECM is therefore an essential component of tissue growth and development. It is a dynamic and even living part of a tissue. Furthermore, the ECM is compositionally heterogeneous, which allows it to specialize in support of specific tissues. In kidneys, highly porous fibrillar networks serve as molecular filters, narrowly sized and regularly tilted orthogonal lattices sustain the convex curvature of the cornea necessary for optical transparency, and the mechanical

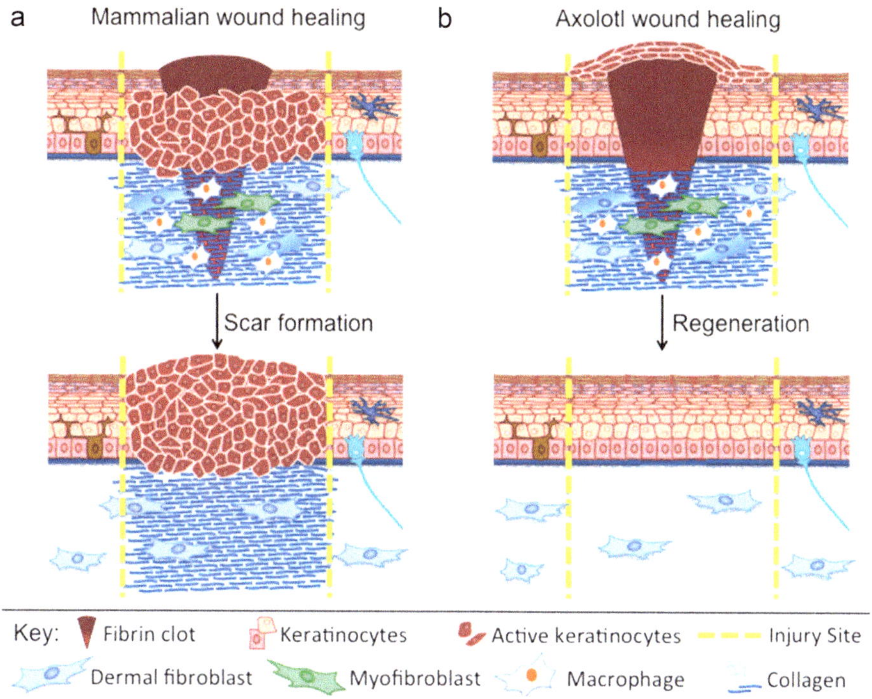

Figure 4.1 Wound healing mechanisms. (a) Mammalian wound healing: a fibrin clot is formed with keratinocytes proliferating under it and fibroblasts entering the wound bed, where they start to express the ECM. Keratinocytes then migrate over to close the wound. Some fibroblasts will differentiate into myofibroblasts, contracting the wound. Scarring leads to a thickened epidermis with excessive and unremodelled ECM. (b) Axolotl wound healing: keratinocytes migrate over the fibrin clot, closing the wound within 24 h, followed by their proliferation with the formation of a wound epidermis; fibroblasts then enter the wound bed, proliferate, and secrete ECM, with some differentiating into myofibroblasts to contract the wound. The cells within the formed dermis continue to remodel the ECM until regeneration is complete. Reproduced from ref. 11 with permission from Elsevier, Copyright 2018.

resilience of the ECM provides elasticity in cartilage and fracture resistance in bone.[15] Such a functional diversity permits and requires significant variability of ECMs belonging to distinct ECM families the functional expressions of which correlate with their composition.

4.2.1 Structural ECM Constituents

Matrices of connective tissues such as cartilage, bone and skin incorporate extensive extracellular connectors and are assembled as polymolecular assemblies of proteoglycans and collagens I and II. These matrices have relatively smaller amounts of growth factors, but are rich in adhesion proteins, entactin and amorphous collagen IV in basement membranes, which help

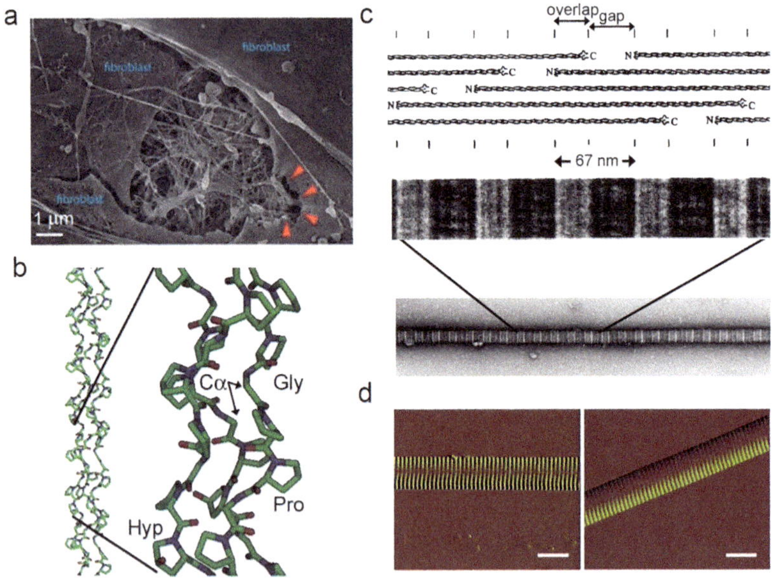

Figure 4.2 Collagen structure. (a) Electron micrograph of a native extracellular matrix deposited by fibroblasts (centre). Red arrows point to the damaged membranes. Reproduced from ref. 13, https://doi.org/10.1016/j.yjsbx.2018.100002, under the terms of the CC BY 4.0 license https://creativecommons.org/licenses/by/4.0/. (b) The collagen triple helix built from three (Pro–Hyp–Gly)$_n$ chains. Arrows indicate the Cα atoms of glycines (PDB 2D3F). (c) Schematic of tropocollagens assembling into five-stranded "early fibrils" with an axial D-periodicity of 67 nm (top) and electron micrographs of a collagen fibril with the characteristic periodic pattern of light and dark bands, and an *N*,*N*-bipolar collagen fibril with tapered ends (bottom). The higher magnification box indicates the 8D-long region of polarity transition. Reproduced from ref. 27 with permission from Elsevier, Copyright 2008. (d) Atomic force microscopy images of collagen fibrils with the characteristic striation-like D-periodicity. Scale bars are 0.5 µm. Reproduced from ref. 28 with permission from the Royal Society of Chemistry.

them to endow boundaries between different tissue types and to mask enzymatically susceptible cryptic sites.

The ECM consists of many individual components communicating with cells at the molecular level. The components help the matrix to amplify and transmit signals between messenger molecules and cell receptors regulating intracellular pathways. The role that the ECM plays in signal transduction and in the regulation of cell behaviour renders it indispensable for cell responses supporting functions ranging from cell adhesion and apoptosis to vascularization and organ morphogenesis.[16]

This property of regulating multiple extra- and intracellular processes is both unique and complex, and emulating it has been a major focus of research towards induced tissue regeneration. The literature describes many approaches on how to achieve this, most of which address the challenges of creating synthetic fibrillar matrices or finding alternative gel-type, viscoelastic materials able to reproduce the biomechanical properties of the ECM. These two related paths are often pursued concomitantly, with a main rationale being in replicating the compositional nature of the native ECM.

The ECM is a complex fibrillar material assembled with nanoscale precision. Collagen is a protein that makes up the bulk of the ECM, accounting for one-quarter of the total protein in the human body. Collagen is also a conserved protein fold characterized by a tripeptide motif, Gly–Xaa–Yaa, where Xaa and Yaa can be any amino acid, but are usually proline (Pro) and hydroxyproline (Hyp), Figure 4.2b. Tandem repeats of the Gly–Pro–Hyp motif, with site-specific insertions of some other amino acids, make up contiguous polypeptide chains 300 nm in length. The chains interdigitate into rod-like bundles 1.5 nm in diameter. One bundle constitutes a right-handed helix comprising three nearly identical left-handed helices. The triple helix is referred to as tropocollagen. Side-chain-free Cα atoms of glycine residues placed at every third position of the motif are oriented towards the interior of the fold, which has no space for any side groups, Figure 4.2b. This is induced and held by the outward orientation of the proline rings. Polyproline helices are very common in proteins and are characterized by the absence of intrachain hydrogen bonding necessary for other oligomerizing elements. Left-handed helices are lowest in energy for polyprolines and are stretched and open to accommodate local conformational inversions caused by glycines. Inclusion of glycines in polyprolines may also be required for derivatization (*e.g.* glycosylation, cross-linking) of collagen fibrils *via* modified amino acids (hydroxyproline, hydroxylysine, allysine), as these were found to be specifically oriented relative to glycines in the triple helix. The folded tropocollagen is the building block of collagen fibres.[17–19]

There exist over 30 genetically encoded collagens (*e.g.* I, II, III, V, IX and XI). All contain a domain with the triple helix conformation, and all are integral components of the ECM. However, only a few are involved in the fibre formation itself.[20,21] As judged by electron microscopy, the fibres appear as morphologically uniform cylinders with a striated surface pattern of light and dark bands that repeat every 67 nm, Figure 4.2c.[22]

The striated pattern is described as an axial D-period and is used as a measure unit of the fibre length.[23] The length of one tropocollagen (300 nm) is approximately 4.5D. Tropocollagens aligned side-by-side in a parallel, regular stagger by multiples of D assemble spontaneously into five-stranded early fibrils of about 90 periods (~7 μm), Figure 4.2c. These are intermediate or seed structures, the further coiled-like associations of which lead to mature fibres hundreds of microns in length. Type I and II collagens are major fibrillar components and are the ones that are synthesized in response to injury.

However, to become functional these fibrils must copolymerize with collagen molecules of other types. For instance, types III and V co-assemble with type I into heterotypic fibrils, whereas type XI determines the structure and assembly of type II fibrils. Type IX, also known as a FACIT (fibril-associated collagens with interrupted triple helices), decorates the surfaces of fibrils.[24] FACIT is a small collagen subfamily of four types, none of which can form fibrils, but each can associate with fibril-forming collagens. In FACIT collagens, triple helix domains are separated by non-collagen (NC) domains, which can be glycosylated and can contain cysteine, methionine or hydroxylated amino acid residues that covalently cross-link and stabilize fibrillar structures. FACIT collagens modify fibres, but also mediate their interactions with cells. For example, collagen IX binds oligomeric matrix proteins such as fibronectin and vitronectin. These proteins interact with cell surface receptors, integrins, which mediate mechanical changes in the matrix. FACIT collagens can also regulate lateral associations of the early fibrils into mature fibres. Similarly, the function of a FACIT-related collagen IV is to generate sheet-like networks in basement membranes, which are arranged by hexameric hubs of two trimeric NC domains, assembled through head-to-head interactions and stabilized by covalent cross-linking of hydroxylysine residues.[25] Other collagens that can be found in various tissues are equally important for the functioning of the ECM.

As a rule, failed or abnormal co-assembly of non-fibrillar collagens with fibrillar types causes tissue disorders. For example, gene defects in collagens IX and XI develop functionally anomalous matrices that lead to degenerative diseases such as osteoarthritis and chondrodisplasia syndromes. Basement membranes lacking collagen IV disintegrate with time, leading to kidney failure, as in Alport syndrome, and reduced production of collagens III and V is the main determinant in the development of intracranial and arterial aneurysms and connective tissue degeneration. The spectrum of diseases caused by collagen mutations is broad, emphasizing the importance of preserving native collagen assembly, Figure 4.2c and d.[26–28]

4.2.2 Fibrillogenesis as the Mechanism of Building the ECM

As a typical representative of scleroproteins, collagen undergoes fibrillogenesis – a spontaneous assembly into extended fibrils. The exclusion of bulk water from the surfaces of individual collagen molecules allows them to aggregate into a highly ordered paracrystalline phase, fibre. The cylindrical shape of this phase is a thermodynamically stable form providing a minimized surface area-to-volume ratio. Protein filaments are ubiquitous structural components of biological systems, ranging from fibrin matrices to intermediate filaments.[29,30] These structures exhibit nanoscale periodicities similar to those observed in collagen fibres – cross-striated surface patterns indicative of a paracrystalline order,[31] which indicates that protein fibre formation obeys general physical principles regardless of the nature and origin of protein building blocks.

Such principles manifest in the polar or directional assembly of the blocks and their lateral associations or bundling. The globular proteins tubulin and actin undergo a directional assembly with the formation of microtubules and microfilaments, respectively,[29,32] whereas the assembly of fibrinogen molecules into fibrin fibrils is initiated at one end by thrombin.[30] Intermediate filaments assembled from vimentins and lamins are rope-like structures of long contiguous α-helices bundled together. Long individual helices are not stable in aqueous media, with bundling stabilizing the assembly. For instance, in intermediate filaments individual helices first coil around each other, forming 2 nm wide rod-like dimers, which then associate in a head-to-tail fashion to form filaments 10 nm in diameter.

The observation of D-periodicity in collagen fibres in addition to other continuous nanometre periodicities in fibril systems points to a unipolar mode of assembly. Collagen fibres tend to taper with carboxyl and amino termini closest to the opposite tips. In some cases, fibrils exhibit a polarity reversal having the same ends.[33,34] Such a reversal results from a switch in molecular orientation in a region of the axial repeat of approximately eight D-periods. Furthermore, collagen fibres of some invertebrates are symmetrically bipolar, which is particularly intriguing as bipolar fibres have uneven diameters, with the thickest parts located at the switch regions. The resulting spindle-shaped fibrils are limited to a few microns, which can be indicative that they represent nucleating structures. Indeed, both unipolar and bipolar fibrils elongate *via* tip fusions into longer fibrils, suggesting a mechanism involving unipolar fibrils as the main building blocks and bipolar fibrils as capping elements. This mechanism gives rise to longer bipolar structures with depleting unipolar reactants. Thus, the ratio of unipolar to bipolar fibrils determines the final composition and the size of a collagen matrix. Although bipolar fibrils cannot fuse, homogeneous mixtures of unipolar fibrils associate into star-like aggregates, with longer fibrils formed by 4D tip overlaps generating the thinnest and mechanically fragile regions.[35] Subsequent thickening of the fusion sites stabilizes the structure, possibly involving lateral associations with tropocollagens. In contrast, small proteoglycans conjugated to the fibrillar surfaces inhibit side-to-side fusions along fibrillar surfaces. As collagen fibres mature, they branch and form networks *via* tip-to-shaft fusions of collagen fibres that result from the interactions of tips with switch regions.

Branching is a common mechanism for the generation of fibrillar networks employed by different, folding-unrelated scleroproteins. Fibre branching renders the ECM three-dimensional and defines its viscoelastic properties. Individual insoluble fibres, when branched, gel and become compatible with the cell environments that they are designed to support. Therefore, understanding branching mechanisms is important for designing biomimetic matrices.

As an exemplar, fibrin fibres assemble from twisted fibrinogen-converted protofibrils that associate laterally by wrapping around each other.[36,37] Protofibrils are half-staggered structures exhibiting a periodicity of 22.5 nm in matured fibres, suggesting surface tension as a limiting factor for the aggregation of protofibrils.[37] Consequently, with most of the fibres forming native clots being straight, displaying no or little curvature, Figure 4.3a, curved and branched fibres are also observed, Figure 4.3b.

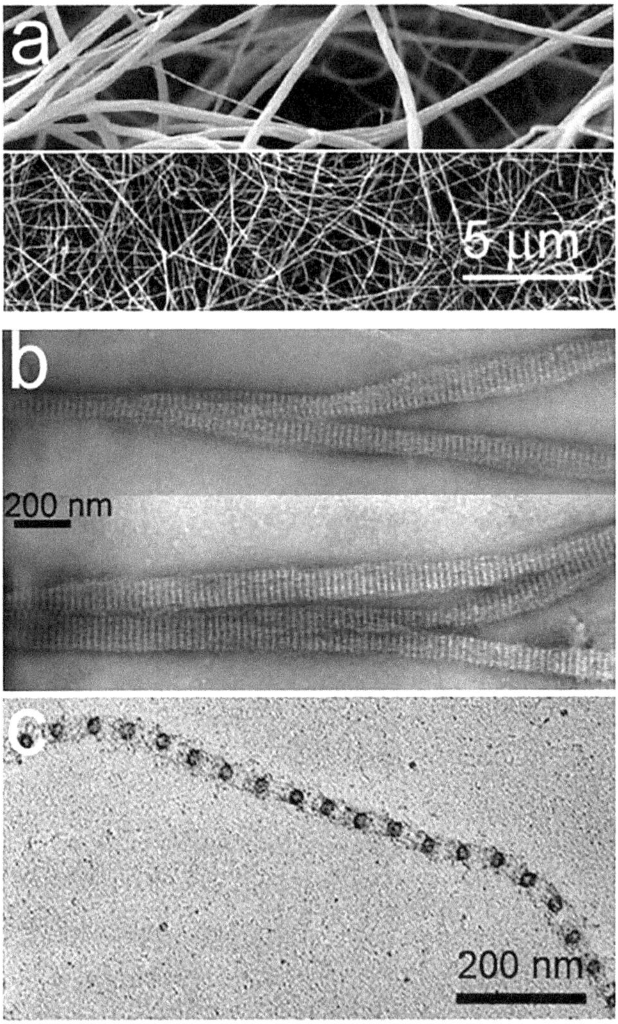

Figure 4.3 Fibrin and fibrillin. (a) Electron micrographs of fibrin clots showing thick fibres with few branching points (top) and thin fibres with many (bottom). (b) Electron micrographs of branched fibres with the characteristic band pattern of 22.5 nm. Reproduced from ref. 36 with permission from Elsevier, Copyright 2004. (c) Electron micrograph of fibrillin-rich beaded structure extracted from human fetal membranes. Reproduced from ref. 41 with permission from Elsevier, Copyright 2016.

Branching points are built up from three fibres and can be separated at various distances depending on the conditions in the polymerization of fibrinogen. The band patterns appear to be perfectly aligned, with two fibres joined with a small angle at almost all branching points, Figure 4.3b. This morphology may indicate a growing fibre that is split into two at an early polymerization stage or branching initiated as a result of the addition of other growing protofibrils.[38]

Noteworthily, the bundling and branching of protofibrils appear to be antagonistic events in fibrin and the preference for one or the other is dictated by the conditions of fibrin assembly: high concentrations of thrombin (initiator of the polymerization of fibrinogen into fibrin) favour branching, leading to dense networks, whereas low concentrations produce much thicker fibres with few branching points.[36]

However, branching is not the only option available for the formation of networks. Other polymerizing proteins that co-assemble into the extracellular matrices of some tissues can also support the formation of intricate networks.

The elastic properties of connective tissues of lung, aorta and skin are attributed to elastin and fibrillin, which are the core building components of elastic fibres. These proteins polymerize independently in a stepwise manner and merge in a final stage, resulting in cross-linked networks.[39]

Fibrillin first assembles into microfibrils with periodicities of 56–60 nm. These regions have a bead-like appearance and are separated by stretched string regions able to uncoil and coil back, thus endowing the fibrils with unique elastic properties, Figure 4.3c.[40]

Elastin assembles into fibrils as tropoelastin, which has two distinctive, alternating domains, insoluble and cross-linking. Tropoelastin molecules are aligned on the preformed microfibrils, which serve as templates or scaffolds for directing the deposition of tropoelastins and subsequent cross-linking.[40,41] Rubber-like matrices capable of responding to an externally applied force by stretching and retracting are the final product of this mode of assembly.

Branching and cross-linking events in fibrillogenesis underpin the physicochemical properties of the matrices expressed in gelation. Virtually all variants of the ECM are hydrogels. This property guides approaches towards the design of synthetic hydrogels for tissue engineering and as research tools for 3D cell culture. The dry protein content of hydrated collagen and fibrin fibres does not exceed 2%, which helps create appreciable porosity within fibrillar networks without compromising their viscoelastic properties. The pore sizes in cell-free hydrogels, however, appear much smaller than the sizes of even the smallest cells. Therefore, protein hydrogels may perform better when premixed with cells prior to cross-linking or branching. Elucidating the relationships between the physical and biological properties of hydrogels guides the development of instructive regeneration. However, reproducing the whole repertoire of these native properties in a particular biomimetic system may not be feasible. Therefore, different approaches must be tried. One popular path is to reconstitute the native ECM. This is not exactly a design approach but pursues a similar aim of morphogenetic adaptation of the ECM, which in most cases starts with the extraction of the native ECM from a biological tissue.

4.2.3 ECM Materials from Biological Sources

Placental, dermis, pericardial, vascular, bone and connective tissues are major sources of collagenous ECMs. A variety of methods are used to extract collagen from these tissues, including acid extraction, high-speed homogenization at high salt concentrations and thermal and vacuum dehydration. These methods have been successfully used to reconstitute collagen gels. As a rule, the materials obtained are composed of non-cross-linked collagens. The mesh sizes of the networks formed are on the cellular length scale, but somewhat larger than in the native state, which can prove useful for studying cell motility and tissue mechanics in healthy and cancerous tissues. For example, metastasizing cells were shown to be able to orchestrate the pericellular remodelling of the ECM by selectively executing fibre breakdown. As the cells move forwards, they align sterically hindering fibres into microtracks of a single cell calibre. Other cells that follow expand these tracks by degrading lateral ECM interfaces, which provides a route towards collective invasion.[42] Complementary to this is the role of the ECM in supporting tissue contraction, which is important for healing processes, as the inability of a pericellular environment to defy contraction forces may lead to pathological contractures. Therefore, synthetic mimetics must exhibit high levels of mechanical strength and compatibility to respond to remodelling by migrating cells. Gel extracts can be adjusted to exhibit such properties by using chemical (glycosylation), enzymatic (transglutaminase) and physical (elevated temperatures) cross-linking or biomineralization that can yield more stress-responsive gels.

Chimera matrices are another effective means of mimicking the biomechanical properties of native tissues. Elastin-like protein networks, reinforced with adjustable collagen fibres, can be engineered into spatially aligned composites or tubular conduits.[43,44] These materials exhibit controllable mechanical responses at the microscale and offer bespoke fabrication platforms for synthetic tissues with predefined and controlled extracellular composition. The polar mode of the collagen assembly can be exploited in the design of polarization-responsive architectures to generate pyro- and piezoelectricity in the axial direction of collagen assembly.[45,46] By employing the collagen polarity, it is possible to generate magnetically oriented gels or achieve the preferentially non-mechanical alignment of fibrillar matrices on non-biological surfaces.[47] Biomaterials with enhanced collagen alignments can be expected to be beneficial for directional tissue growth as in neurite and vascular outgrowth. Admittedly, however, collagen extracts are not devoid of shortcomings, which are mainly related to immunogenicity and disease transmission. A solution to such risks is the use of recombinant and chemically synthesized collagens.[48,49]

Extractions of fibrin and elastin materials from biological sources are also practised: cryoprecipitation of blood plasma is used to obtain fibrin, while elastin is reconstituted from human or animal tissues using acidic or hot alkali treatments. With their remarkable adhesive properties, autologous fibrin materials find use as tissue adhesives and glues for surgery. Traditional applications of fibrin include skin grafts, knee arthroplasty, vascular, intestinal and colonic anastomosis and seroma prevention. Fibrin sealants are used in craniomaxillofacial and orthopaedic surgeries as carriers for the slow release of demineralized bone, bone morphogenetic proteins and growth factors. The rubber-like properties of elastin are exploited for the regeneration of tissues that require resilience and elasticity, *e.g.* arteries, lung, ligaments and skin. The deposition of elastin sheets on cross-linked hyaluronan gels advances the construction of functional composites that can mimic complex multilayered structures of aortic valves, the elastin-rich ventricularis or viscoelastic spongiosa. Biological extracts, their semisynthetic derivatives and synthetic analogues have established an impressive track record in biomedical applications.[50,51]

From the diversity of these applications in both clinical and laboratory settings, it is clear that elastin, fibrin and collagen materials have matured to offer a range of properties necessary for most tissue types. All three materials define the mechanical and structural responsiveness of the tissues and provide sustained support to tissue growth and repair. This role, however, is not exclusive to collagen, fibrin and elastin assemblies. Other biopolymers exhibit similar viscoelastic and lubricant properties, *e.g.* the mucopolysaccharide hyaluronic acid, a major element of connective, epithelial and neural tissues, and synovial fluid.

Irrespective of the differences between these biopolymers and the tissues they support, the key driver behind the aim of designing extracellular matrices is to reproduce the viscoelastic and gel-like properties of the native extracellular matrices.[52] One of the major hurdles on the path to the successful design of fibrous matrices is the mesh size of fibrillar networks, which in artificial designs appears to be much smaller than the diameters of most cells. At least partly, this limitation can be addressed by using large-pore gels reconstituted from raw collagen-rich extracts. However, such extracts have unresolved drawbacks which are not characteristic of synthetic materials. While synthetic analogues of the ECM hold promise for an ultimate solution to the problem, biopolymers from non-animal sources present an alternative route to gel-like materials. Alginates are one such non-animal biopolymer.[53]

Alginates are polysaccharide components of natural gums – materials able to increase viscosity in a concentration-disproportional manner. The brown algae are a major source of this material. Alginates are random anionic composites of β-D-mannuronic (M) acid and its epimer, α-L-guluronic (G) acid, which are fused together *via* glycosidic linkages into linear copolymeric chains, Figure 4.4a.

M and G blocks are coupled into three main types of monomeric blocks – MM, GG and GM. The distribution of the blocks in polymer chains is the major determinant of the physicochemical properties of alginates. The average square end-to-end distance per uronate residue is approximately two times larger for G than for M, suggesting that alginates with more G are stiffer and that their properties can be controlled. Further, MM and GG blocks adopt different conformations: polyMM prefer linear or threefold left-handed helices, whereas polyGG fold into stiffer twofold helices stabilized by intramolecular hydrogen bonds between and within GG blocks. GG blocks rigidify polymer chains and G-rich alginates present denser materials, which is in contrast to amorphous MM-rich polymers. Alternating

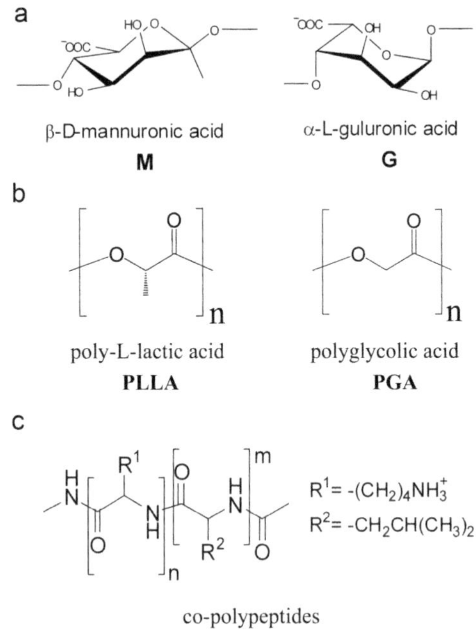

Figure 4.4 Chemical structures of hydrogel-forming polymers.

MG blocks makes polymers more flexible, resulting in less dense gels with superior elasticity. PolyMG adopt conformations with hydrogen bonds between the carboxyl group of M and both hydroxyls of the following G. This binding increases conformational freedom, leading to greater flexibility, which in alginate polymers increases in the order GG < MM < MG. Therefore, the simple conversion of MM or GM to MG improves the elasticity of the polymer:[54] M-rich segments are more elastic, but do not contribute to cross-linking, which is necessary for gelation. To gel, alginate polymers need divalent cations, *e.g.* Ca^{2+}, Ba^{2+} or Sr^{2+}. In the presence of the cations, <2% aqueous solutions of alginates gel instantaneously. Stronger gels reach an equilibrium between stiffness and elasticity, which allows further increases in flexibility linearly correlated with the number of intermolecular cross-links formed by divalent cations. GG blocks interact with divalent cations forming egg-box-like clusters between adjacent polymer chains, which enables gel-supporting cross-linked networks to be formed. MM blocks prefer ribbon conformations, which use carboxylic groups to chelate multivalent cations in an intramolecular fashion, discriminating gel formation in favour of amorphous liquid materials.[55]

Such a tuneable behaviour of alginates can suit different applications. Moreover, the ratio of G which is produced from M can be programmed during the alginate biosynthesis. The strength and elasticity of the required gel can thus be made responsive by tuning the molecular weight, content and distribution of G/M blocks in the polymer chains. In native environments, alginates use low molecular weight polymers with high GG contents to meet the demand for high alginate concentrations by algae species growing in raw seas. Similar reasoning has been used to engineer alginate-based materials.[53] Alginate gels are readily produced and effectively operate under physiological conditions. The physicochemical and mechanical properties of these materials offer significant control over key performance parameters such as drug release, cell binding, biodegradation and erosion. Relationships between the structure of alginate gels and these parameters have been elucidated with the aim of improving biomedical compatibility.[53,56]

A preliminary focus has been on the mechanical properties of the gels and their role in *in situ* tissue growth. This is of direct relevance to the search for desirable gel forms for regenerative medicine – tissue implants.[53] Such formulations must be able to maintain physical integrity while supporting assigned functions during the formation of a new tissue.[57] Controlled degradation of alginate gels is one of the key performance factors in this regard. Alginates, given their non-animal origin, degrade slowly *in vivo*. Their degradation is more difficult to control than that of protein matrices containing specific signal sequences, which are degraded by matrix metalloproteinases, thus allowing for matrix remodelling. Therefore, achieving better control over the degradation of alginates continues to be a major research focus for the development of alginate-based implants.[53,57] Alginate hydrolysis was shown to be possible *via* periodate-mediated cleavage of carbon–carbon bonds in uronate residues.[58] Degradation of the polymer occurred within 9 days, with the rates of degradation tuned by external stimuli, such as changes in pH and temperature. Partially oxidized alginates retained their ability to bind calcium ions and cross-link, forming gels. By adjusting the molecular weight distribution of such gel-competent alginates, it was possible to control elasticity. Increased fractions of low molecular weight alginates maintained similar rates of elasticity to those of the high molecular weight gels. This approach gave more rapidly degrading hydrogels that were more receptive to transplanted bone marrow cells in the formation of new bone tissues.[59] Alternative approaches involving other modifications can also be employed to control the degradation of alginates. For example, the incorporation of cell-adhesion motifs in the gels was shown to enhance the mechanical properties of alginate gels at the expense of stronger interactions between modified alginates and cells.[60]

Other modifications involve the use of different biocompatible polymers: cationic polyethylenimine to reinforce alginates, gelatine to promote rapid cross-linking of oxidized alginates, or chitosan to allow polyion alginate complexations. The potential and diversity of cross-linking methods to improve the performance of alginate gels are significant. However, it is the alginate itself that remains a limiting factor for the progress of alginate-based therapies. In this regard, synthetic polymers with the useful properties of alginate gels may provide more straightforward access to biodegradable gels.

4.3 Mimicking ECM by Design

4.3.1 Polymer Mimetics of the ECM

Much research has been carried out to develop synthetic polymers as biomimetic scaffolds.[61] Synthetic scaffolds can be produced at any scale to common specifications. The mechanical, physicochemical and degradation properties of polymeric matrices are relatively straightforward to control and tailor. Their use is not associated with known immunogenic responses and disease transmissions, which are common for biological extracts. In principle, any non-toxic polymer can be used to produce cell-supporting scaffolds and water-soluble polymers can be formulated to support gelation. Synthetic polymers used as regenerative templates range from polydioxanones to ceramics. Among others, polyesters, such as poly(glycolic acid) (PGA), poly(L-lactic acid) (PLLA) and their copolymers (PLGA), are perhaps the best studied and most commonly used polymers in medicine, Figure 4.4b.

One of the most attractive features of these materials is that they degrade to natural metabolites, eventually converting to carbon dioxide and water. The degradation of the polymers does not involve enzymes as ester bonds are susceptible to hydrolysis, the rate of which can be predefined by specifying the ratio of the monomer types.

These polymers are also rigid thermoplastics with high crystallinity (~50%) and can be manufactured into different shapes. Their high sensitivity to hydrolytic degradation requires careful control of the processing conditions and implies the prefabrication of implants. As a consequence, the types of materials produced from the polymers are predominantly porous solids and foams. Chemically inert materials able to give stable gels in water are of high demand in medicine, which is mainly attributed to hydrogels possessing properties desirable for implants and atypical of other types of biomedical materials.

Biocompatible and biodegradable hydrogels are relatively straightforward to construct. Methods for their degradation and the type of degradation by-products can be rationally assigned through the selection of safe building precursors. Hydrogels exhibit a low interfacial free energy with body fluids, which prevents unspecific adhesion and adsorption of cells, bacteria and biopolymers. Hydrogels are elastic materials, which makes them capable of minimizing tension and irritation to surrounding tissue.

Elastic hydrogels are usually made from water-soluble polymers at lower or upper critical solution temperatures. Block copolymers of poly(ethylene glycol) (PEG) and poly(propylene glycol) are common examples. Similarly to alginates, these polymers gel as a result of a liquid–gel phase transition. Whereas alginates require divalent cations to form gels, polymeric glycols undergo thermoreversible gelation. However, at physiological temperatures the polymers are also prone to diffusion into the liquid phase, which creates stability issues for implants based on synthetic gels. Cross-linking strategies used to stabilize natural elastomers are also efficient for synthetic polymers.

Toxicology requirements for cross-linked synthetic polymers are stricter, particularly for approaches that require the premixing of cells with the precursor components of the gel. The use of organic media and the toxicity of reactive molecular species generated after cross-linking contribute to undesired side effects. All these factors are considered when selecting polymeric candidates and methods for their polymerization and stabilization. Photopolymerization, transglutaminase-catalysed condensation and biointeractive Michael-type addition reactions appear to provide the necessary balance between the efficacy of production and biocompatibility. These and similar strategies have already been used to advance live vaccine delivery, cell encapsulation, cartilage and bone repair, wound healing and the like.[62–70]

Unlike matrices reconstituted from biological extracts, synthetic polymers are bottom-up materials and can be assembled in tissue culture from low molecular weight precursors. Thus, not only can these materials encapsulate cells upon gelation but they can also be made responsive to changes in their environment to mimic the inherent property of the native ECM, matrix remodelling. This mechanism of matrix degradation and resynthesis is a complex process that remains overlooked in many matrix designs. Matrix restructuring enables cell invasion and migration and accommodates cell proliferation, which permits efficient processes to allow matrix turnover in a controllable and reproducible manner. The degradation of the native ECM is regulated by cells that it encapsulates. The process can be emulated in engineered synthetic hydrogels by a combination of cross-linking metalloproteinase substrates and pendant adhesion sites.[70] The substrates assembled into branched polymeric networks interacting with cells *via* cell-adhesion RGD-based peptide motifs can follow cell-mediated proteolytic degradation and further remodelling into cell-secreted matrices at injury sites. The synthetic method included addition reactions between conjugated unsaturated ends of PEG chains and thiol-containing peptides, to give rise to bioactive PEG networks upon mixing the components under physiological conditions. Swelling measurements on the material obtained revealed that the network mesh sizes, taken as the distance between consecutive cross-links, were considerably below (<50 nm) cell sizes and the length scales of cellular processes. This was used to assess the sensitivity of the gels to metalloproteinases as cell migration in this case is limited to proteolytic degradation. Fibroblast outgrowth was found to be hindered for gels lacking metalloproteinase substrates. Cell migration was also dependent on cell-adhesion motifs (RGD ligand), with no cell invasion taking place in the networks lacking the ligand.

Similar strategies find use in the fabrication of tissue organoids.[71] A typical organoid is a 3D multicellular *in vitro* model of a tissue.[72] The use of organoids necessitates the development of synthetic extracellular matrices, which, unlike biological systems, are not predisposed to a particular cell type, nor they can contribute to disease transmission. Organoids exhibit high stiffness sensitivity, prompting the use of cross-linked hydrogel precursors allowing for controlled biodegradation.[71] Dynamic cross-linking proves to be indispensable for engineering mechanically dynamic matrices, which after partial degradation should remodel without losing their mechanical properties. Exemplified by applications in intestinal organoid cultures, such smart matrices maintained their mechanical properties over several days, and permitted the expansion and differentiation of stem cells and subsequent formation of specialized intestinal organoids, Figure 4.5a and b.[73,74]

Smart matrices should also be able to display self-control in maintaining their mechanical and physicochemical properties under external stimuli such as pH, temperature or electric field.[75] Self-recovering gels are expected to overcome the main limitations of current hydrogel designs, one of which is a relatively long time needed for hydrogels to swell or to activate under an external stimulus and to deactivate once the stimulus is reversed.[76] All of the characteristic properties of a given hydrogel – elasticity, degradation, stability and mechanical strength – can be controlled by the rate with which the hydrogel can recover. Synthetic methods used to produce hydrogels impact on this directly. Traditional polymerization methods lack the capacity to specify sequence, length, composition and conformational preferences of resulting materials. Polydispersity and structural heterogeneity of hydrogels are also common. Therefore, hybrid production methods have been pursued to circumvent these problems. An effective strategy is to combine the ease of polymer synthesis with the accuracy of

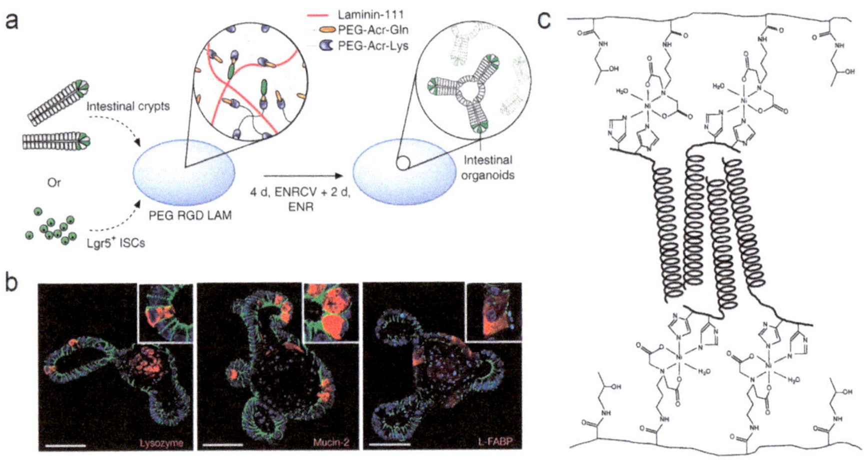

Figure 4.5 Engineered hydrogels for organoids. (a) Intestinal organoids are formed by embedding intestinal stem cells into mechanically dynamic matrices functionalized with RGD and laminin-111. The matrices provide stiffness support for the expansion of the stem cells and with time soften, permitting cell differentiation and morphogenesis. Reproduced from ref. 74 with permission from Springer Nature, Copyright 2017. (b) Fluorescence micrographs of intestinal organoids containing differentiated intestinal cells. Scale bars are 50 μm. Reproduced from ref. 73 with permission from Springer Nature, Copyright 2016. (c) Assembly of protein–polymer hydrogels mediated by metal complexation. Reproduced from ref. 79 with permission from Springer Nature, Copyright 1999.

molecular recognition. Several groups have proposed different models of such an approach, focusing on the self-assembly of block copolymers. Early efforts included the construction of stimuli-sensitive and reversible polymer hydrogels, using triblock copolymers of synthetic and biopolymers.[77] Self-assembling coiled-coil sequences conjugated with synthetic polymer chains can be used to provide interchain polymer interactions that give rise to polymolecular networks. Coiled coils are ubiquitous protein folding motifs, which are often used in nanoscale biodesigns as controllable, non-covalent cross-links. The motifs are bundles of two or more individual helices brought together *via* complementary interactions. In block copolymer networks, coiled-coil domains swell in solution once formed, while the solubility of the networks is guaranteed by the polyelectrolyte domains of the conjugates. Individually, peptide sequences remain unstructured and soluble in water, but when folded together into coiled coils they formed stiff, rod-like cross-links, supporting reversible swelling, rather than precipitation, and tuneable gel–solution transitions. Further studies proved this to be important for controlling hydrogel erosion. Hydrogels, which are more resistant to erosion, were produced by tuning the strength of protein–protein interactions and controlling the overall network topology.[78]

A somewhat reversed approach, in which coiled-coil domains formed independently of the polymer chains, provided similar effects. In this case, the domains were used as pendant groups interacting with methacrylamide polymer chains in a non-covalent manner. The domain sequences incorporated histidine tags, which, like methacrylamide, can complex Ni^{2+} ions. Thus, hydrogels are assembled by mixing chelating methacrylamide copolymers charged with Ni^{2+} and the coiled-coil domains, Figure 4.5c.[79] This design permits the use of other protein domains provided that they contain histidine tags. Proteins with different properties can introduce additional performance qualities, such as responsive sensitivity to temperature. Human cardiac titin domains used as cross-linking domains in such graft copolymers demonstrated a temperature-dependent increase in swelling above their melting temperature (58 °C), which was in contrast to chemically cross-linked hydrogels lacking such responsiveness.[76] Several features make these hydrogels distinct: metal-mediated complexation of proteins containing affinity tags, the use of divalent ions for protein-mediated stabilization of hydrogels, and applications atypical of naturally occurring hydrogels, *e.g.* an ability for *in vivo* monitoring using magnetic resonance imaging.

The concept of smart hydrogels extends to engineering systems with rapid on–off responses. For instance, diblock copolypeptide amphiphiles composed of poly(L-lysine) or poly(L-glutamic acid) and poly(L-leucine), poly(L-valine) or poly(D/L-leucine) as the hydrophilic and hydrophobic components, respectively, produce rapidly recovering hydrogels, Figure 4.4c.[80] This design is notable in that copolypeptides of identical compositions, albeit of statistically random sequences, do not produce hydrogels. However, when synthesized as narrowly distributed low molecular weight polymers, the reported amphiphiles associated into hydrogels at low concentrations. Furthermore, the hydrogels were found to maintain their mechanical strength at high temperatures and were able to recover rapidly after applied stress. The latter correlated with structural parameters of the polymeric chains and in particular with their ability to fold under gelation conditions.

Hydrophobic domains can adopt regular conformations, which are insoluble in water, namely poly(L-leucine) forms rod-like α-helices, whereas poly(L-valine) favours crystalline β-sheets.[81] By capitalizing on these differences, it was possible to reveal empirical correlations for conformation-specific gelation: the tendency to gel was higher for helical segments. Gelation also depends on the porosity of the forming macromolecular meshes. Networks of higher porosity gel at faster rates. Therefore, the fast gelation observed for these gels can be attributed to high porosity as the peptide amphiphiles were found to concentrate within a gel matrix leaving liquid voids polymer free. The combination of the two features (rapid gel recovery and high porosity) are important for applications that require the controlled release of encapsulated materials or drugs. One way to meet this requirement is to use cleavable bonds, which can be introduced into the polymer backbone or cross-linking domains, and which can be broken under physiological conditions at whim.[82] Controlling the degradation and recovery rates of hydrogels allows tissue growth and delivery of therapeutic materials to be monitored, although an important factor to consider in hydrogel designs is the biocompatibility of their degradation products, which are not necessarily the same as building-block precursors. However, the range of unique properties that hydrogels offer and the relative ease with which these can be achieved outweigh the shortcomings of hydrogel designs in stimulating new designs.

4.3.2 Fibrous Mimetics of the ECM

To this end, non-fibrous mimetics and hydrogels have been discussed. These designs can emulate the physical properties of the ECM, including gelation, viscoelasticity, mechanical and tensile strength and, in some cases, responsiveness to external stimuli. The native ECMs are, however, fibrous protein materials the physical properties of which are defined at the subcellular level. Molecular recognition, folding and self-assembly patterns and also the nanoscale mechanics of cell–matrix contacts provide instructive cues for tissue development and constitute response triggers to accommodate changes in cellular environments. Overall, these considerations stimulate the search for the nanostructured fibrillar designs of artificial ECM analogues.

Provided that all of the main requirements concerning biosafety are met, an ideal synthetic hydrogel may arise from any molecular class. There are indeed no limitations on what chemistry can be used to deliver a successful candidate. This creates ample opportunities for novel chemistries and fabrication methods and may give a material of exceptional quality and efficacy with functions that are not characteristic of the native matrices. However, with time, success stories become fewer as more modifications and optimizations are required to approach the performance level of native tissues. Therefore, all similarities with native matrices should be tested and exploited. In this light, nanoscale designs are expected to reveal the missing link between an optimal synthetic matrix and the fibrillar nature of the native ECM. New possibilities emerge following the progress made in the development of high-resolution measurement methodologies such as cryo-electron microscopy, atomic force microscopy and chemical and correlative imaging techniques, which help to gain more insight into the function and architecture of the ECM with increasingly high accuracy and precision. Electron microscopy these days is routinely used to analyse the morphological features of fibrillar materials irrespective of the fabrication methods used. Manufacturing methodologies may vary, but validation strategies remain the same as they essentially assess the same design form. X-ray diffraction techniques provide a suitable probe to confirm the paracrystalline nature of protein fibres, the biological role of which is still poorly understood. High-resolution microscopy helped to relate the crystallinity of protein fibres to their surface patterns, such as D-periodicities of collagen and fibrin fibrils, which are taken as guiding parameters for new designs. Emerging measurement approaches promise to reach beyond these capabilities. Notable examples are the development of *in situ* electron microscopy, super-resolution microscopy and atomic force microscopy capabilities, which can measure molecular and cellular processes in native aqueous environments in real time.[83–86] Such capabilities, while being developed, encourage designs that mimic the very processes of ECM formation. Figure 4.6 shows a model of

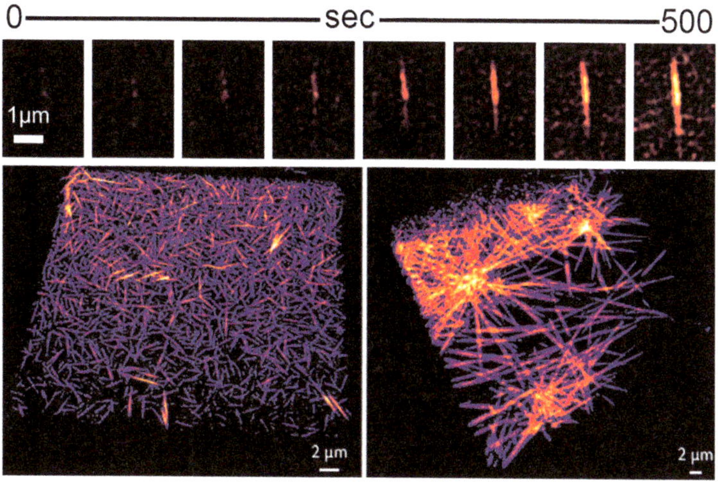

Figure 4.6 *In situ* imaging of protein fibre formation. Fluorescence micrographs of designer protein fibres obtained using structured illumination super-resolution microscopy. Top: self-assembly of an individual filament measured over time. Bottom: two different morphologies of filaments assembled from one peptide component (left) and two peptide components (right). Reproduced from ref. 85 with permission from American Chemical Society, Copyright 2017.

a bidirectional fibre assembly imaged using super-resolution microscopy.[85] The model assembles from a *de novo* polypeptide and emulates the longitudinal assembly of collagen fibrils. Unlike collagen fibres, however, it is designed as a homogeneous system in terms of the morphology, length and width of individual fibres. The addition of a complementary peptide redirects the assembly into star-shaped materials, suggesting branching nucleated in star hubs, in a manner similar to that of native collagen and fibrin fibres, Figure 4.6.

The native collagenous ECM results from the co-assembly of different components based on a very specific folding element. Following these established modes of assembly would seem to be an obvious approach to access a synthetic collagen analogue. Surprisingly, however, there have been few attempts towards synthetic collagen-based assemblies. This can partly be explained by the challenges facing the chemical synthesis of proline- and glycine-rich sequences, limiting the success of collagen-sized building blocks. Still, a number of impactful designs have helped initiate the experimental elaboration of collagen mimetics. A popular approach has been to use short collagen sequences which self-assemble by forming a stagger with complementary overhangs.[87] The core principle of these designs is that extended triple helices could be generated by a poly- or supramolecular elongation of shorter peptides provided that they support strong intermolecular interactions. A combination of regular staggering and forced intermolecular interactions can be used to design heterotrimeric self-complementary peptides. Synthesized as classical (Gly–Pro–Hyp) tandem repeats, the peptides were laterally linked upon oxidation of cysteine residues incorporated into the opposite ends of the chains to ensure the stagger. The linked peptides had 12- or 13-residue complementary collagen overhangs each, which assembled into larger aggregates.

In another collagen triple helix design, similar assumptions were used.[49] The design was chosen to incorporate an auxiliary capacity that would be able to limit considerably the promiscuity of potential interactions, as collagen assembly is not mediated by a code or a recognition pattern. Covalent bonds formed by a pair of disulfide bonds were used to tether the strands of collagen fragments and also to offset the strands in setting the register. Consistent with the design, an intramolecular triple helix between different segments produced collagen-like fibrillar structures.[49]

Electron and atomic force microscopy data confirmed that the morphology and diameter of the resulting assemblies strongly resembled those of native collagens, although a broad distribution of fibre lengths from 30 to >400 nm was observed. This tendency proved to be common in collagen designs and can be attributed to impaired stability of designed assemblies compared with native collagens. The problem could be addressed through chemical modifications of polypeptide backbones that in some cases led to the stabilization of the triple helix and stronger interhelical interactions.[88] However, designing more efficient polypeptide assemblies that at a later stage can be adapted to recombinant technology would be more attractive in this regard.

A solution was proposed in the design of triple helix protomers that assembled into D-periodic microfibres reminiscent of the native fibrillar collagen.[89] Protomers are sequences of three different (Xaa–Yaa–Gly) domains, with a sequence containing Pro–Hyp–Gly repeats used as a core hydrophobic domain, which was coupled to two oppositely charged sequences of similar sizes. Electrostatic interactions between arginine residues of one charged domain (cationic) and glutamate residues of the other (anionic) ensured the linearly aligned and axially staggered assembly of triple helices. An assumption made in this design was that such complementary, electrostatic interactions are reinforced by hydrogen bonding, which are set by the core segment.

An external trigger is often necessary for the formation of ordered collagen structures. Under benign conditions and at low salt concentrations, amorphous fibres are observed. In contrast, thermally annealed protomers can give smooth fibrils hundreds of nanometres in length and tens of nanometres in diameter.[89] Longer post-melt incubations give uniform fibres, micrometres long and 70 nm in diameter, with a characteristic transverse banding. The axial D-period observed in protomer designs was ~18 nm and formed by 63 residues, in comparison with the 67 nm periodicity of tropocollagens containing over 1000 residues. The apparent discrepancy between the peptide length and the found banding distance suggests that the nucleating unit of the assembly may be a lateral oligomer of protomers, the staggered bundling of which is defined by regions, albeit periodic, but of varied packing and alternating charge density. This observation bears relevance to fibre designs based on other folding motifs. Other, varied compositions of the collagen (Gly–Pro–Hyp)$_n$ pattern can be introduced in synthetic collagens to permit the assembly of collagen heterotrimers, which in some instances proved to produce materials with properties reminiscent of those of native collagen materials. A promising paradigm was described as the use of oppositely charged strands in collagen triple helices.[90] This can be achieved by introducing cationic (lysines) and anionic (aspartates) amino acid residues into collagen amino acid triplets, Figure 4.7. Stable conformers of charged residues are formed and are accessible for electrostatic interactions, which sets contiguous electrostatic networks between the strands, thereby ensuring the assembly of extensive collagen fibrillar structures and hydrogels.[91,92]

Hydrogels continue to represent the most advanced class of tissue-supporting biodesign materials. Polymer-based versions are cross-linked three-dimensional networks that encapsulate water throughout the volume, creating a non-fluid colloidal material. The networks constitute the continuous phase of the hydrogel, which can be tailored for better control over gelation. However, once a polymer has formed a hydrogel, its degree of polymerization cannot be altered. By contrast, cross-linking can be applied at any stage of hydrogel formation, allowing for a broad range of mechanical properties ranging from elastomer to resin. Cell self-renewal and differentiation correlate with the physical properties of the niches that they occupy.

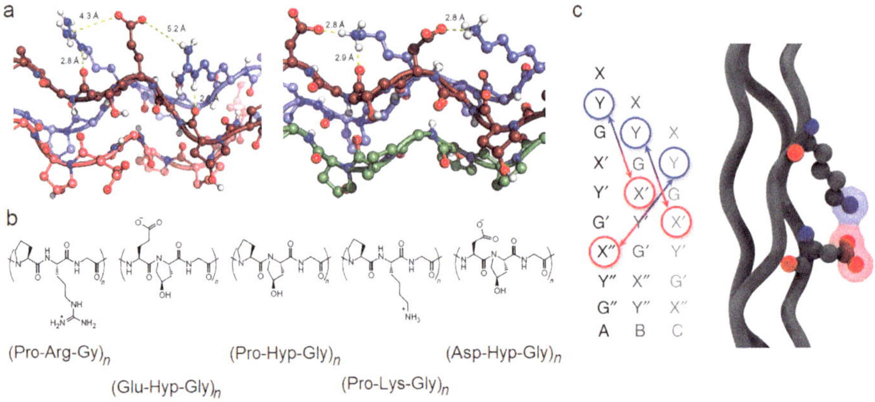

Figure 4.7 Fibrillar collagen mimetics. (a) Models of electrostatic interactions for Arg–Glu (left) and Lys–Asp (right) in collagen triple helices. Two stable Lys conformers are formed in Lys–Asp, allowing hydrogen bonding to Asp. (b) Chemical structures of amino acid triplets in the design. Reproduced from ref. 90 with permission from Springer Nature, Copyright 2011. (c) Left: relative position of Lys–Asp axial interactions which are necessary for the triple helix formation. Right: molecular representation of the interactions. Reproduced from ref. 91 with permission from Springer Nature, Copyright 2021.

With the tensile strength resembling that of developing tissues, hydrogels impact most on applications in surgery and wound healing. In these applications, it is important to provide hydrogel materials with elastic properties matching tissue stiffness from $E \approx 1$ kPa (Young's modulus) for soft tissues, such as brain or fat, to 50 kPa for hard tissues, such as bone.[93] Tissue elasticity can direct the controlled release of drugs, growth factors and nutrients that are readily encapsulated into hydrogels.[94]

First attempts at sourcing fibrous hydrogels from tissues rich in the native ECM exposed compositional heterogeneity characteristic of biological extracts, batch-to-batch variations and poor utility for translational research, which stimulated synthetic designs exhibiting nanoscale order. The use of purified collagen, alginate and fibrin materials revealed similar tendencies of lacking reproducibility and control, but also helped to reveal major factors necessary for proteinogenic hydrogels. An underpinning factor can be defined as the formation of nanostructured fibres mimicking those of native collagen or fibrin materials. Therefore, before long it became clear that the shortest and perhaps most reliable route to fibrous mimetics is to emulate the process of fibrillogenesis. Self-assembled protein or peptide fibres assume the role of polymeric chains in hydrogel scaffolds. Covalent cross-linking is of little value here because such fibres are not individual, but bundled polypeptide chains: linking two chains in the bundle or between bundles does not give stable interconnections. Fibrous cross-links and networks provide a more practical solution. These can be statistically random overlaps of individual fibres, branching fibres or interconnected structures.

In all cases, porosity, defined as water-filled spaces between fibre chains, appears to be important and can be enabled using net-like architectures with pores extending several microns in diameter.[95]

Using the analogy of polymer chemistry, fibre assembly is a supramolecular polymerization of polypeptide blocks the size and nature of which are of less importance.[96] The rationale of chemical polymerization and cross-linking in polymers manifests in supramolecular polymerization and networking in protein fibre designs, which are not limited to specific sequence lengths or folding elements. In essence, any folding motif capable of self-assembly can be designed to assemble into fibre: short synthetic collagens and collagen mimetics, α-helices and β-strands are all fit for the purpose.[95–100]

4.3.3 Non-collagenous Assembly Mimetics of the ECM

Interfacial hydrophobic interactions in conjunction with oppositely charged domains have been exploited in the supramolecular polymerization of designed peptides into fibres. Indeed, fibre systems assembled from non-collagen sequences have been used, ranging from β-pleated sheets and α-helical coiled coils to swapping domains.[101–109] Such designs share two main requirements which prove to be necessary for fibre formation.

First, a self-assembling sequence has to be complementary or self-complementary. Complementarity ensures the formation of homo- or heterooligomers and is readily reproduced by design. However, it does not guarantee self-assembly.

Therefore, second, a self-assembling sequence should incorporate a property that allows it to propagate. An effective approach is to alternate charges within the sequence. First demonstrated[102] in the assembly of β-structured fibrillar membranes and matrices, the application of this principle has been demonstrated in many other designs. To give an example, a self-complementary sequence comprising alternating hydrophobic and polar residues – the classical β-structure pattern – segregate the side chains of the two amino acid types into two surfaces – hydrophobic and polar or charged. The hydrophobic surface of moderately hydrophobic alanines, together with the polar face comprising alternating basic arginines and acidic aspartates, generates a peptide repeat, RAD or RADA. An alternative version used a reverse of lysines and glutamates correspondingly, that is, EAK. The two peptide designs were termed "modulus two", reflecting the arrangement of charged residues on the ionic side of the folded β-sheet – when folded, two positively charged residues would be followed by two negatively charged aspartates, and *vice versa*. Other ionic moduli can also be produced. These are distinguished numerically according to a modulus number which is correlated with the number of alternating basic and acidic amino acid residues. For example, − − + + − − + + consistent with the described modulus would be numbered II, whereas I would stand for − + − + − + and − − − − + + + + − − − − + + + + for IV. A number of different modulus versions have been described.[110,111]

This format allows for a mode of assembly which involves the formation of β-sheets *via* intermolecular hydrogen bonding, alanine and charged residues shaping hydrophobic interfaces and intermolecular ionic bonds, and the staggered association of individual peptides. Peptide monomers assemble as staggered β-sheets setting up the assembly of extended protofilaments, which, by undergoing side-by-side association, give macroscopic fibrous membranes and tapes.

The model is consistent with spectroscopic data supporting a strong preference for β-sheet formation, an almost quantitative estimation of the β-content for both designs, and with the results of electron microscopy, which revealed a strictly hierarchical assembly of folded peptides. Under various conditions, such designs could produce distinctive microscopic morphologies from membranous structures to tape-like assemblies. Protofilaments 10–20 nm in diameter assembled into fibrous networks with a relatively narrow distribution of pores of 50–100 nm. The designed sequences may be too short for appreciable folding and are significantly shorter compared with the components of native matrices. However, once formed, all of the generated materials were stable,[112] which re-emphasizes the efficacy of using short self-assembling sequences for ECM mimicry.

Furthermore, architectures based on alternately spaced alanines of axially shifted sequences are characteristic for other native fibrous proteins. For example, silk proteins have fully extended backbones forming β-sheets with staggered alanines. With the smallest side chains, alanines arrange the densest packing of hydrophobic faces, creating contiguous threads of β-sheets to form macroscopic protofilaments spontaneously and irreversibly.[113]

The ratios of small amino acids in fibrous proteins tend to be high, which proves to be universal for the monomer packing in fibrous assemblies. In β-structured proteins such as keratins or fibroins, and also in others represented by different folding elements, notably collagens and elastins, the incorporation of glycines or alanines is faithfully reproducible over the entire sequence lengths. This considerably relieves steric constraints during the formation of hydrogen bonds between neighbouring polypeptide chains, which can otherwise be imposed by larger side chains preventing lateral association and alignment. Similarly, in collagens and elastins steric hindrance is undesirable owing to the limited interior space within helical filaments. Such a prevalence for small and chemically inert amino acids can serve as a structural signature for proteins whose preferences over chemical specificity lie in the closest possible packing of building blocks. This may also be the reason for the incorporation of specific functional domains within the scaffolding sequences of structural proteins or the existence of specialist proteins within separate structural families. Interestingly, the "modulus two" design combines both functions in one sequence, providing an excellent example of space economy.

The assembly of either modulus peptide is both spontaneous and instantaneous in a range of aqueous buffers, including cell culture media. This is compounded by observations that the RAD motif used in one of the peptides may mediate integrin-dependent cell adhesion in a manner similar to that of the fibronectin RGD.[114] Therefore, the RAD-based designs can express two functions necessary for artificial extracellular matrices – formation of microscopic networks and cell adhesion – which was confirmed experimentally.[115]

The described principle of electrostatic complementarity in conjunction with the staggered assembly of hydrogen-bonded β-pleated structures is applicable to other designs, folding elements and assembly patterns. Of practical interest are the similarities between α-helices and β-strands. The number of residues per turn for β-strands is lower than that for α-helices, 2 *versus* 3.6. This renders β-strands stretched helices. The rise per residue for β-strands is higher than that of an α-helix, 3.3 Å against 1.5 Å, which is thermodynamically unfavourable for a structure with such parameters to form autonomous structures. Therefore, singular β-strands are rare and tend to be stabilized by intermolecular hydrogen bonds.

In marked contrast, α-helices are autonomous structures. Their folding does not depend on oligomerization, nor does it prevent helices from associating laterally as would be required for a staggered assembly,[116] which implies the formation of at least a dimeric structure with two helices axially shifted with respect to each other.[117] Helices meet this criterion by folding into coiled coils.[118–120]

However, staggered coiled coils are not normally found in natural systems. Known examples are very few, and include tropomyosin and intermediate filaments, which use abrupt modes in their assembly.[121,122] Most of the known coiled-coil structures prefer lateral association and rely upon oligomerization by bundling. The preference is not entirely clear as the staggered type of assembly would seem to be more effective in generating extended fibrillar forms of the same dimensional parameters. One explanation may come from a sticky-ended coiled-coil design, which was conceived as a test for the ability of short helical peptides to assemble longitudinally into an infinite coiled coil, Figure 4.8a.[105,109]

Instead, two complementary peptides designed to co-assemble with the formation of a staggered heterodimer with oppositely charged overhangs or "sticky ends" propagated into an extended self-assembling fibre.[105] The fibres appeared thicker than expected for a coiled-coil dimer, 2 nm *versus* 50 nm, indicating the lateral association of coiled-coil dimers:[123] the extended coiled coils initially form 2 nm thick rod-like protofibrils, which then bundle up into thickened fibres, Figure 4.8a. X-ray diffraction in combination with electron microscopy revealed a packing in axial register in a 3D hexagonal lattice of size ~2 nm with 4 nm periodicity along the fibre axis, which equates to the length of one folded peptide block.[123] The measured parameters suggest that the C-termini of preceding peptide blocks abut the N-termini of the succeeding blocks. The termini are oppositely charged and hence may contribute stabilizing interactions into the structure, which creates an opportunity for chemoselective ligation within each non-covalent dimeric coiled-coil strand. Given the model of assembly, in which tens of thousands of peptides align with their N- and C-termini abutting, the formation of microns-long

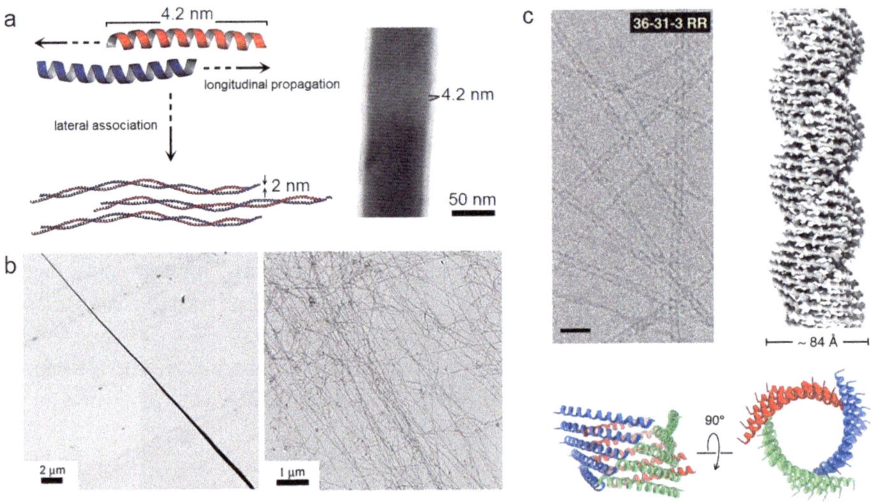

Figure 4.8 Helical filamentous assemblies. (a) Two peptide blocks, each 4 nm in length, fold into a sticky-ended coiled-coil dimer (PDB 2ZTA entry rendered by PyMol). The dimer propagates in both axial directions, forming coiled-coil protofilaments that associate laterally into mature fibres – shown are a model (left) and an electron micrograph (right). The resultant fibres exhibit a periodic striated pattern with striation separated at 4 nm distances, corresponding to the length of the peptide monomers. (b) Electron micrographs of a resultant fibre (left) and ligated fibres (right), in which one peptide block was converted into a thioester, Adapted from ref. 124 with permission from American Chemical Society, Copyright 2007. (c) Cryo-electron microscopy image (left) and its reconstruction (right) of designed peptide filaments, and (below) an atomic model of the assembly showing side (left) and top (right) views. Scale bar is 20 nm. Adapted from ref. 134, https://doi.org/10.1038/s41467-020-20689-w, under the terms of the CC BY 4.0 license https://creativecommons.org/licenses/by/4.0/.

polypeptide chains should be expected. With one of the two peptide blocks functionalized with a C-terminal thioester moiety, the blocks became ligated into extended, covalent polypeptide chains, which proved to be ≥10 µm long and estimated to be ≥3 MDa in mass.[124] The ligation impacted on the thickening of the original assembly generating thinner and much more flexible structures, suggesting that a level of control of fibre thickening can be introduced.[124]

This is intriguing and finds analogies in naturally occurring fibrillar α-helical structures. For example, vimentins, desmins and some keratins allocate highly conserved arginine clusters in their sequences to help regulate inter- and intrahelical electrostatic interactions, allowing control of the stiffness and stability of the filaments formed.[125,126] The conspicuous thinning observed in designed fibres tends to correlate with decreases in their stiffness and stability, resulting in partially branched or disintegrated (shorter) structures. The lowest observable cut-off in thickness, however, does not reach 2 nm (the diameter of a coiled-coil dimer), which suggests that there must be a concomitant mechanism involving both lateral and axial control over the assembly in order to produce uniformly sized 2 nm filaments.[127,128] Although such a mechanism remains to be found, ultimately, the property of controlled thickening can be replicated in building blocks themselves. Solvent-exposed positions in coiled-coil sequences are obvious mutation sites to endow fibre assemblies with stimuli-responsive properties. These positions determine the nature and strength of contacts between individual coiled-coil strands and subsequently impact on the morphological and physical characteristics of the resulting materials, which can be tailored to support specific functions, such as gelation. <Thermoresponsive coiled-coil fibres can be designed to form gels upon cooling with a storage modulus (G') of 1000 Pa. As with other gel systems, gelation is accompanied by the formation of interwoven fibres which can support cell growth and proliferation.[129]

When it comes to thickening, however, natural systems appear to have already been optimized. For example, more than 60 different sequences in the family of intermediate filament proteins assemble into 10 nm thick filaments – with the same resulting morphology irrespective of the type of the protein and its function and origin.[125] Point mutations in these sequences underpin inheritable diseases, which are often associated with the loss of the native filament structure.[130,131] This structure–function dependence re-emphasizes the importance of careful sequence designs to construct homogeneous and reproducible α-helical fibrillar structures. Attempts are being made to devise empirical rules for designing sequences assembling into filamentous structures with controlled thicknesses. Correlations are commonly derived from the polarity of solvent-exposed residues in coiled-coil sequences, coiled-coil oligomerization states and the length of peptide sequences, which also define the length of the filaments.[132–135] Reports proposing unifying models for filamentous assembly are also relatively frequent. However, these consider particular sequences for particular forms of filaments, and focus primarily on accurate ultrastructural analysis of the designs rather than on providing a unifying mechanism or design rationale. In other words, a rationale explaining which mutations in any coiled-coil sequence can convert one thickness to another with 1 Å precision remains to be established.

4.3.4 Monomorphic Fibre Mimetics *versus* Matrix Formation

One of the early attempts to devise a general model consisted in the construction of helical blocks having the same heptad repeat sequence, ensuring the formation of the same interhclical interactions, which become shifted by multiples of the heptad repeat to generate an axial stagger.[132] The stagger between adjacent helices must be equivalent to an integral number of the repeats or a value divisible by one heptad. Importantly, the i and $(i + n)$ helices of an n-stranded filament should not overlap upon the completion of one turn. The shift in this case would not be more than one residue, which is to provide an adequate space for head-to-tail packing of α-helices. Larger shifts would generate larger gaps between abutting ends of the peptide blocks, which can then lead to irreparable aberrations in the assembly. The most favourable helical arrangements to form n-stranded filaments would obey a general formula $(n \times 7 - 1)$, that is, helices composed of 20, 27 and 34 residues would promote the assembly of three-, four- and five-stranded coiled-coil filaments, respectively.

A dimer assembled from a 13-mer sequence would be the simplest design. However, this sequence falls short of two heptad repeats to form a persistent assembly. Therefore, a 34-residue sequence was chosen to exemplify the rationale. Consistent with the prediction, this sequence assembled into five-stranded coiled-coil filaments 2.5 nm in diameter, although the formation of fibrillar structures was not exclusive. Spherical assemblies 10–15 nm in diameter were also observed, suggesting that control over thickness carries a cost of stability. An approach to address this problem is to regularly space bulky hydrophobic residues, such as tryptophan, in a coiled-coil stagger to control interfacial interactions. The repeat positioning of such interfacial residues in the formed coiled coils defined the packing of resultant filaments, which proved to arrest thickening by maturation.[133] However, in most such cases, existing and emerging, structurally distinct filament sub-populations arising from the same assembly mode are common. As it stands, a predictable design of monomorphic filament structures continues to represent a significant challenge. An effective strategy has recently been proposed to control lateral associations of peptide protofilaments using an arginine clasp motif, RxxxR. The motif was shown to mediate cohesive interactions between protofilaments exhibiting a *de novo* cross-α supramolecular architecture, Figure 4.8c.[134] Similarly to arginine clusters in vimentins and desmins,[126] the clasp motif appeared to regulate interhelical interactions.[134] Unlike in natural systems, the motif was site-specifically allocated in heptad repeats to interact with the C-terminal residues within adjacent helices, creating highly periodic interfacial interactions, which translated into the persistent and regular stacking of individual helices perpendicular to the axis of a growing filament. This design rationale represents a conjecture that the axial and lateral interfaces of the filament should be conserved, which allows the thickness of the resultant filaments to be correlated with the length of peptide components from the first arginine in the motif to the C-terminus of the adjacent helix. The subsequent computational refining of the design allowed monomorphic filamentous structures with a highly conserved ultrastructure and thickness to be generated, Figure 4.8c. A generic computational strategy was devised to correlate filament diameters with the number of helical repeat units in the monomer. This strategy exploited multiple intermonomer interfaces to generate multilayered filaments templated on a cyclic symmetry. The number of lateral layers in the filament was shown to retain the interaction interfaces but the diameter of the filaments changed linearly with the number of repeat units in the protomer, *i.e.* three, four, five or six repeats giving filaments of increasing diameters.[136]

Other strategies used to set up and control filamentous forms have explored interplays between different types of non-covalent contacts, *e.g.* hydrophobic, electrostatic and π–π stacking of aromatic residues. Small hydrophobic patches can be introduced in the termini of peptide blocks to favour their axial propagation, with electrostatic interactions supporting lateral assembly.[137] For example, a two-residue stagger in the heptad repeats, unlike a one-residue shift in *n*-stranded filaments discussed earlier, was designed to generate two hydrophobic surfaces in the helix, which were oriented at ~200° with respect to one another. Such a configuration leads to a third level of supercoiling and promotes the formation of dimers with hydrophobic "sticky" clasps. This approach is different from alternating ionic and hydrophobic faces in the β-sheet moduli or oppositely charged "sticky ends" of coiled-coil fibres, in that the clasps enhance the contribution of the hydrophobic effect initiating directional polymerization, which can be fostered by cross-linking residues, such as cysteines, or activators of the hydrophobic effect (*e.g.* glycerol). The resultant structures tend to be shorter than in other designs, wispy and more polymorphic, which comes at a cost of promoting unspecific associations within the filaments. Monomers with radially functionalized aromatic moieties capable of π–π stacking interactions, *e.g.* bipyridines, can provide a better level of control over the reversibility of helical fibrillogenesis, with the resultant fibres exhibiting regular banding patterns characteristic of near-crystalline forms.[138] Strictly, these approaches cannot limit lateral associations to a particular thickness. Interestingly, however, thickening may prove to be consistent with the fact that bends and kinks in filamentous structures appear to be more common for thinner filaments. Although no clear correlations are known between the frequency of such morphological features and filament widths, irregularly split fibrils happen to form in thin filaments.

The challenge of fibre thickening in helical fibres can be viewed as an acid test for designer systems in that it allows structure–function predictability in engineering highly ordered and persistent nanoscale systems to be demonstrated. The progress achieved in the construction of monomorphic filaments is significant. However, an outstanding question remains as to whether there is a clear need for monomorphic designs. Are precisely controlled size, morphology and ultrastructure important? On the one hand, and as discussed earlier, sequence mutations in natural protein filaments may lead to diseases, suggesting that control over the structure must be maintained. However, is it this resultant structure or the mode of its assembly that needs to be conserved? Natural assemblies, from collagen fibres to α-helical intermediate filaments, do not tend to be limited to a dominating length. Their assembly modes permit structural polymorphism, which helps them accommodate discontinuities in the assembly, probably to be able to respond to the physical cues in their environment.

Branching, networking, splitting and intertwining are characteristic manifestations of morphological deviations, which are necessary for biological function. However, synthetic designs demonstrating these features remain in their infancy.

In part, this can be explained by the fact that nanoscale order may rigidify fibre assembly, making it less responsive to environmental cues. Regular, striated banding patterns are common for natural fibres, such as fibrin and collagen fibres or intermediate filaments, Figures 4.2c,d, 4.3b and 4.9.

Similar patterns could indeed be reproduced in artificial helical and collagen designs, revealing generic parameters in the packing of peptide building blocks in the assembled fibre. Periodic banding patterns are found in almost every design type tried, *e.g.* collagen mimetics and coiled-coil fibres. Moreover, they generally appear to run across the filament axis, Figures 4.8a and 4.10.[139,140]

As an exemplar, the spacings between striations in the sticky-ended coiled-coil design measured ~4.2 nm, which closely matches the distance covered by either of the two peptide monomers (28 residues each) in the folded coiled coil (*i.e.* 28×0.148 nm = 4.144 nm), Figure 4.8a. Individual striations persist along the fibre axis and run across the fibre width, suggesting that the peptide monomers must be in register across the widths regardless of their sizes. Consistent with this, extended monomers comprising 35 amino acids, which span ~5.2 nm, in fibres gave striated spacings of ~5.2 nm.[123]

This design demonstrates the feasibility of controlling structural features within a few amino acid residues. The structural responsiveness of helical fibres can be tuned by peptides of various lengths and different topologies,

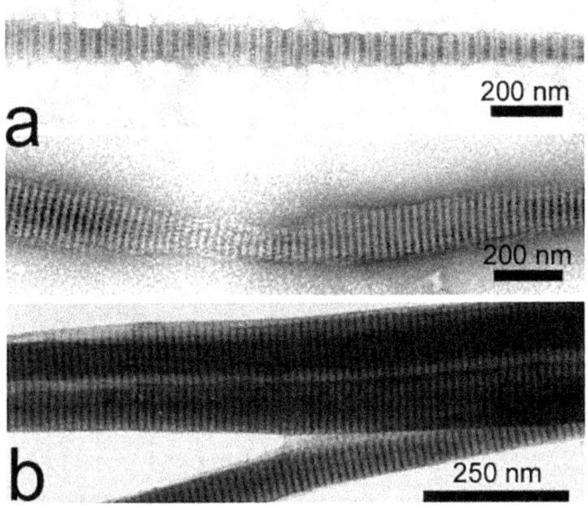

Figure 4.9 Striated banding patterns in natural helical filamentous assemblies. (a) Lamin-derived intermediate filaments. Reproduced from ref. 122 with permission from Elsevier, Copyright 2003. (b) Microtubule-associated fibres assembled from SF-assemblin. Reproduced from ref. 139 with permission from Elsevier, Copyright 1998.

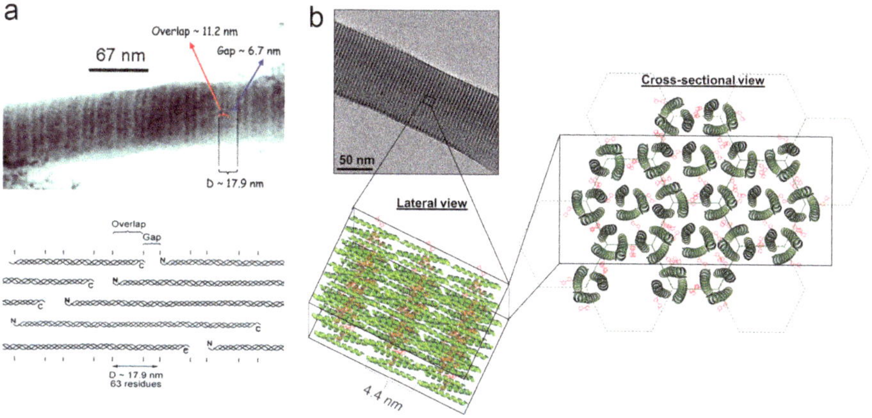

Figure 4.10 Striated banding patterns in designer filamentous assemblies. (a) Staggered assembly of triple helix collagen mimetic peptides into collagen filaments with a D-periodic banding pattern of 18 nm. Each D-period encompasses a "gap" zone and an "overlap" zone. The axial spacing of residues in the peptide monomers is 0.286 nm, which makes a D-period extend over 63 residues. Reproduced from ref. 89 with permission from American Chemical Society, Copyright 2007. (b) Electron micrograph of a coiled-coil fibre assembled from trimeric coiled-coil strands radially functionalized with bipyridine moieties stabilizing inter-strand lateral interactions. The packing of a coiled-coil trimer is indicated within the fibre (lateral view), and also the cross-sectional view (right) shows hexagonal close packing of coiled-coil trimers. Reproduced from ref. 138 with permission from American Chemical Society, Copyright 2018.

which can considerably facilitate the generation of fibre matrices and also their functionalization with biologically relevant recognition motifs – a logical route to synthetic ECM analogues.[141]

In this regard, it is important to elucidate whether the natural solutions to ECM formation, such as fibrillar branching and networking, can be applied to helical fibres, and what the limits are for modifying the straight fibrillar form. There are over 30 different types of collagen involved in the construction of the ECM. Each of these plays a unique role in the assembly of collagen matrices, which suggests that specific building blocks can be designed to shape fibres into any form. Such special building blocks must be complementary to the assembly in a manner similar to that of FACIT and fibre-forming collagens. They must be able to co-assemble with the self-assembling blocks without arresting the assembly or leading to fibre thickening, but with redirecting fibrillogenesis longitudinally with the formation of contiguous morphologies of branching and networking. In this case, fibres themselves become building blocks for more complex hierarchical structures, *i.e.* matrices. The driving force behind the formation of distinguishable matrix-like morphologies is the quantitative co-assembly of special building blocks or modifiers with fibre-forming building blocks, in addition to the kinetic accumulation of the modifiers in the growing fibre. Achieving this feat proved to be possible for helical fibres, Figure 4.11.[142–144]

In a series of studies, it was shown that modifiers, which are topologically distinct from the fibre-forming peptide blocks, reshape the morphology of growing fibres. The modifiers were designed from the two fibre-forming peptide blocks, each of which is split into two building blocks, thus giving rise to four units or sub-blocks, Figure 4.11a. The units are then recoupled in various combinations as orthogonal, non-linear constructs in a head-to-head or tail-to-tail fashion depending on the orientation of the original sequences. Although permitted to some extent, no amino acid mutations are required by this rationale. The modifiers redirect an otherwise straight, unidirectional fibre assembly resulting in non-linear fibres. Depending on the nature of combined units, a preferential or exclusive microscopic feature is introduced. New morphological forms including kinks, splits, branches, segments or interconnections can be selectively assigned to an individual fibre by a particular modifier, Figure 4.11b.

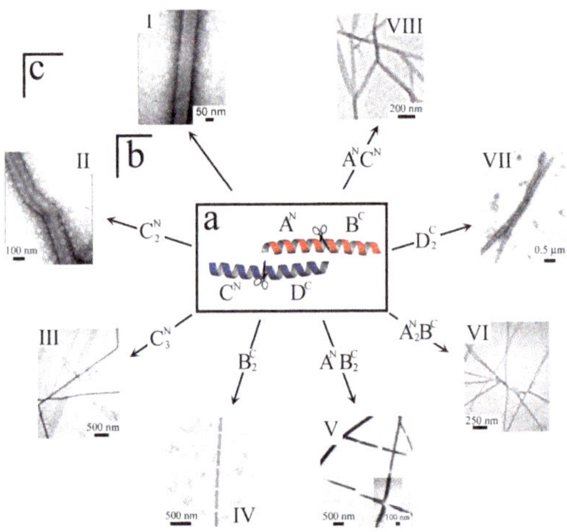

Figure 4.11 Programming the morphology of helical fibres. (a) Schematic of splitting two fibre-forming peptide building blocks into four subunits, labelled A–D, used in the construction of special building blocks or modifiers. Superscripts denote the nature of the free termini, N or C. (b) Modifiers co-assemble with the main blocks, with numerical subscripts indicating the number of subunits (if >1) used in one conjugate. (c) Electron micrographs of various fibre assemblies: (I) standard, (II) kinked with short distances between the two kinks, (III) kinked with long distances between the kinks, (IV) linear and segmented, (V) branching and segmented, (VI) hyperbranched, (VII) split and (VIII) networked. Adapted from ref. 142 and 144 with permission from American Chemical Society, Copyright 2005, and Springer Nature, Copyright 2003.

This empirical algorithm also allows the main architecture to be extended. For instance, it is possible to adjust or tune the distances between kinks or vary the branching density by incorporating subtle changes into the modifiers, Figure 4.11c. Control over such properties, which constitute the breadth of matrix-dimensional parameters and the source of their variabilities, is a challenging task. In particular, this is the case considering that the ECM is a living material undergoing remodelling and exhibiting self-healing, reassembly and cell-instructing properties.

Biodesigns capable of gelation and fibre formation may not be sufficient to provide an effective extracellular support, let alone a fully reciprocal instructive environment. Therefore, other attempts are being made to approximate the morphological characteristics of the ECM beyond the shape of individual fibres. It should be noted here that fibre-shaping modifiers alone do not assemble and are non-linear, orthogonal topologies. Non-linear building blocks, such as branched peptides, can be made self-complementary to produce fibrils that are similar to those produced by linear sequences.[145] However, orthogonal constructs are less attractive for fibre formation owing to the complex synthesis required to make them, while the resulting structures appear to be amorphous and less regular, which is common for promiscuous assemblies.[146] The latter, however, can be enhanced rather than suppressed to set up an arbitrary assembly of ECM-like materials. This mode of matrix formation is deliberately promiscuous and supports fibre formation which is structured at the nanoscale but propagates at higher length scales, from micro-to millimetres.[147]

A conspicuous topology to demonstrate this strategy is a cyclized peptide sequence assembling into fibrillar hyper-branched matrices, which are free of directionality constraints, span microscopic dimensions and can be made biologically differential.[148] This design uses two complementary coiled-coil heptad repeats, anionic and cationic, arranged into an asymmetric unit comprising two domains. One domain (D1) is composed of two anionic heptads and one cationic heptad, while the arrangement of the other domain (D2) is reversed, which gives a split of 2 + 1/1 + 2 pattern setting up various heptad overlaps between domains, Figure 4.12a. Each overlap is probable in any direction with respect to the plane of the cyclic backbone, and the orientation of the domains in the block is antiparallel to ensure interactions between different blocks and not within the same block. The assembly is indiscriminate and encompasses several assembly patterns that can redirect at any point, which introduces a cooperative knot-like propagation of the blocks into mesoscopic fibrillar nets, Figure 4.12b and c. The nets were demonstrated to support cell adhesion and proliferation, which was associated with enhanced formation of filopodia – cell protrusions, which are characteristic of transient adhesions of actively migrating cells that help cells probe their substratum and environmental cues, Figure 4.12d. Therefore, the enhanced filopodia formation suggests enhanced cell–matrix contacts and apparent cellular responses to the physical features of the assembled nets.[149] While supporting human cells, the material efficiently resisted bacterial

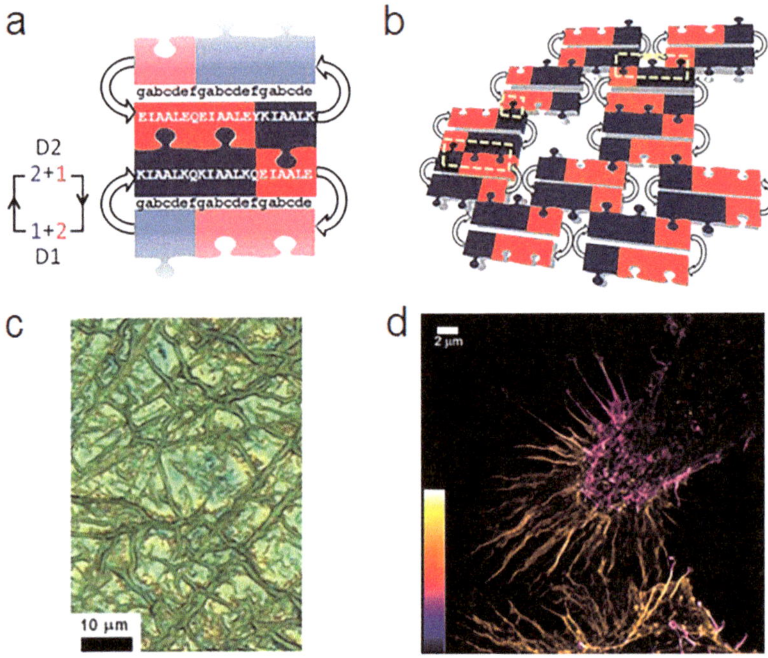

Figure 4.12 Arbitrary assembly of ECM-like fibrillar nets. (a) Schematic representation of an arbitrary self-assembly topology with a 2 + 1 asymmetric unit of coiled-coil heptads in two domains, D1 and D2, and its assembly. One puzzle piece denotes one heptad. Heptad repeats designated "gabcdef" are shown above the linear sequences highlighted in the blocks. The arrows indicate the N-to-C directionality of the sequences. Each arrow is a triglycyl linker. (b) Schematic representation of the assembly with one-, two- and three-heptad overlaps highlighted by dashed yellow lines. (c) An optical micrograph of assembled protein ECM-like nets. Reproduced from ref. 148 with permission from American Chemical Society, Copyright 2014. (d) A fluorescence micrograph showing filopodia formation on a human osteoblast grown on the nets. The image is a colour-coded depth projection distinguishing cytoskeleton actinin lamellipodia projections (violet) and filopodia extending from lamellipodia (orange). The depth colour scale is 1 µm. Reproduced from ref. 149 with permission from the Royal Society of Chemistry.

colonization over several days. This function is realized by the D2 domain, the two-heptad cationic stretch of which has a binding affinity to bacterial membranes.[148] The antimicrobial activity of an individual D2 domain is negligible but is sufficiently significant to disrupt bacterial membranes when amplified in the nets.

Since these assemblies are intrinsically biocompatible and can be made reversible, the approach is particularly attractive for engineering matrices capable of remodelling in cellular environments in response to chemical and physical cues, closely emulating the characteristics of naturally occurring ECM systems.

4.3.5 Minimalist Fibrous Gelators for an Artificial ECM

However, in order to function, these self-shaping or "smart" scaffolds require the incorporation of bioactive groups. Natural extracellular matrices display cell-adhesion motifs, ligands or epitopes on fibre surfaces. Therefore, a logical solution to mimic the ECM functionally is to decorate fibres with biomolecular recognition moieties. Similarly to the design rationale of structural modifiers, this can be done by modifying fibre-forming building blocks with specialist moieties. As with the modifiers, the resulting conjugates must also be compatible with the assembly without arresting or affecting it. Although most of the modified blocks become packed inside the fibre, those that form the fibre exterior would provide sufficient coverage of bioactive groups for recognition by proteins and cellular receptors. In this format, functional motifs act as baits by capturing partner proteins from the solution phase and recruiting them on the fibre surfaces,[150] or as mediators of matrix–cell interactions supporting cell differentiation.[151]

The interactions that such bioactive recruiters introduce can be modelled for fibre systems, which are based on any folding elements since the surface chemistry of peptides is rich enough to support any relevant modifications. A minimalist nanofibre design is based on primary peptide amphiphiles, which assemble into β-pleated micellar filaments, Figure 4.13a and b.[152] The amphiphiles are conjugates of hydrophobic alkyl chains, which become buried in the core of the forming micelles, and polar groups, which can be modified to carry cell-adhesion motifs, which become displayed on micellar surfaces at nearly van der Waals densities.[151]

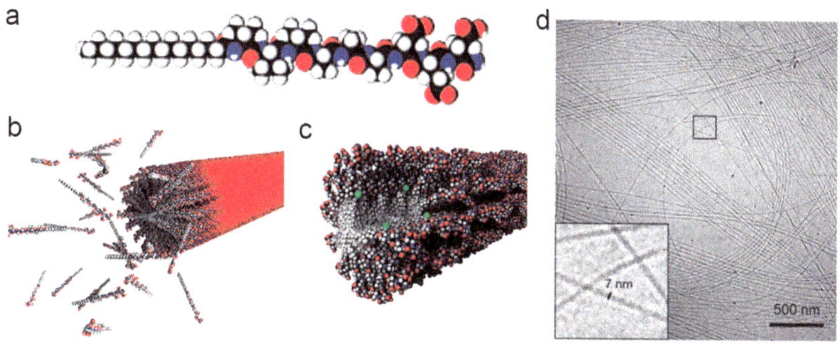

Figure 4.13 Self-assembling peptide amphiphiles. (a) Molecular model of a primary peptide amphiphile and (b) its assembly into a nanofibre micelle. Reproduced from ref. 152 with permission from American Chemical Society, Copyright 2017. (c) Molecular model of a nanofibre assembled from different peptide amphiphile blocks. (d) Electron micrograph of the assembly showing high aspect ratio nanofibre micelles with diameters of 7 nm. Reproduced from ref. 153 with permission from Springer Nature, Copyright 2014.

Despite the apparent simplicity of these designs, primary amphiphiles can incorporate virtually any bioactive motif, label or tag,[153] while sharing the same physical and biological properties with the native ECM, including nanoscale order, fibrous morphology and the ability to support cell adhesion, growth and differentiation.[152] Primary amphiphile scaffolds can be rationally designed to encapsulate progenitor cells[151] and to deliver morphogenic factors to restore bone[154] and enhance burn wound healing[155] and muscle stem cell transplantation,[156] and demonstrate potential as an injectable therapy for haemorrhage.[157]

As an exemplar, amphiphile blocks of this type were designed to assemble into uniformly thin fibres (7–8 nm wide) displaying neurite-sprouting laminin epitopes.[151] Upon assembly in cell suspensions, the fibres appeared as highly hydrated 3D fibrillar networks (>99.5 wt% water), which were rapidly converted to gels embedding cells. Cell-adhesion epitopes present on the fibre surfaces were at densities that proved to be considerably higher than those typically found for the native ECM. The densities can be decreased or controlled using varying ratios of biofunctional amphiphiles, *i.e.* those that are modified with bioactive epitopes, and blank amphiphiles, *i.e.* those that do not incorporate the epitopes. Although a critical density of epitopes for cell binding has yet to be established, van der Waals densities are amply sufficient to support cell signalling.

It is also possible to use different amphiphiles, each carrying a specific moiety, to co-assemble into nanoscale fibres with a functionally diverse corona of bioactive epitopes and labelling tags, Figure 4.13c and d.[153] Given the internal assembly dynamics of these nanofibre scaffolds, contacts between them and cells are also reciprocal in that both the scaffolds and the cells that they encapsulate instruct the dynamics and rate of cell growth and differentiation. The application of decorating fibre surfaces with bioactives has achieved most progress in relation to functionalization with cell-adhesion or homing motifs, aimed at promoting cell and tissue development.[158–161] Other strategies use similar principles of co-incorporation of modifiers or bioactives into fibrous scaffolds. In most cases these are realized *via* co-assembly relying on the complementarity and reversibility of molecular recognition patterns, and in some cases a disruptive, physical means is applied such as strong affinity of collagen mimetic peptides to type I collagen under controlled thermal conditions.

As discussed earlier, peptides designed to mimic collagens are relatively short, but by their nature are prone to form collagen triple helices, which allows them to bind to partially denatured collagen fibres by associating with thermally labile and disentangled domains.[162] The incorporation of these peptides into native and denatured collagens may provide an elegant route to mimicking non-fibrous FACIT collagens which could function as individual decorating units for the collagen fibrils. By inserting into degraded or unfolded collagen chains, collagen-hybridizing peptides (CHPs) can be used to pinpoint and monitor collagen breakdown in various tissues. When tagged with a contrast agent, the peptides permit *in situ* imaging of collagen remodelling during pathological and normal tissue processes, including cartilage breakdown during osteoarthritis or conversion during bone development and collagen degradation in heart tissue following myocardial infarction, Figure 4.14.[163]

Attempts to replicate the morphological and physical characteristics of the ECM, rather than its chemistry, eliminate restrictions on chemistries and molecular topologies that can be applied to fabricate mimetic architectures. This permits any ideas that are progressive enough to mitigate the problem of creating a functional ECM mimetic. The highlighted concepts of ionic alternation, longitudinal staggering and tubular micellization successfully adapted for fibrillogenesis and matrix formation continue to be developed with other functional applications in mind.

The design of complementary β-structured peptide moduli can be tailored to the formation of folding-responsive hydrogels, the assembly of which is triggered by the controlled release of divalent metal ions at set temperatures or in response to exposure to near-infrared light.[164] At low concentrations, the designed peptide folds into an α-helix, but undergoes an α-to-β transition upon release of the ions, leading to the formation of β-sheet fibrillar networks, which gel. Similarly, complementary staggers used in helical designs rely on the transition from a random coil to an α-helix, which can be templated on different peptide topologies, including dendritic geometries. Irrespective of the topology used, the high helical propensity of helical designs is retained in the assembly, from fibrillogenesis to the formation of bundles and matrices, which is consistent with the presumption that no restrictions are imposed on a fibre-like assembly by a particular topology.

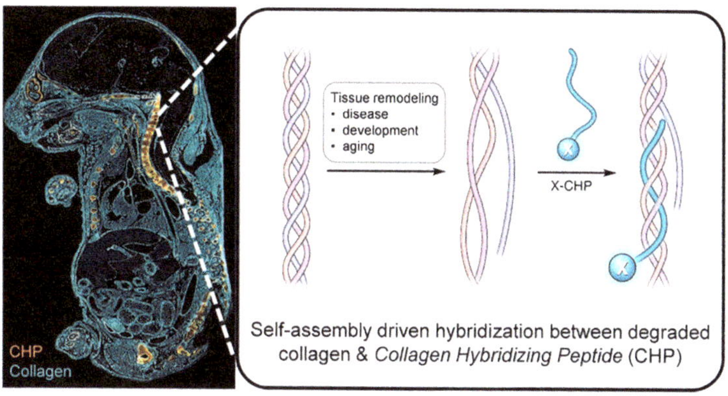

Figure 4.14 Collagen-hybridizing peptides. Left: collagen remodelling during bone development visualized in a mouse embryo using collagen-hybridizing peptides (yellow) incorporating into degrading collagen. Right: schematic of a collagen-hybridizing strand labelled with a fluorescent tag (X), which incorporates into a damaged collagen and hybridizes with it, forming a collagen triple helix. The tag modification enables the location of the damaged sites to be visualized. Reproduced from ref. 163 with permission from American Chemical Society, Copyright 2017.

Other minimalist topologies explore the phenomenon of 3D domain swapping – a naturally occurring mechanism employed by proteins to oligomerize. The mechanism involves exchanging identical structural domains between protein monomers.[165] During the process, folded monomers intertwine, with most of their intramolecular interactions preserved in a resulting swapped structure. An up–down–down topology of *de novo* helices can be used to adopt this mechanism for constructing extended fibrillar structures.[107] An assembly unit is a three-stranded bundle formed between two copies of the same helical hairpin, Figure 4.15a. Helical I and II domains of the hairpin associate with a third domain, III′, of another copy of the hairpin. Remaining I′/II′ and III domains overhang to assemble with other copies, Figure 4.15b. Consistent with this design, microscopic filaments are formed, with the assembly units oriented along the main fibre axis. Interestingly, as in the three- and five-stranded coiled-coil designs, these filaments tend to bundle rather than fuse or mature into thickened fibres.

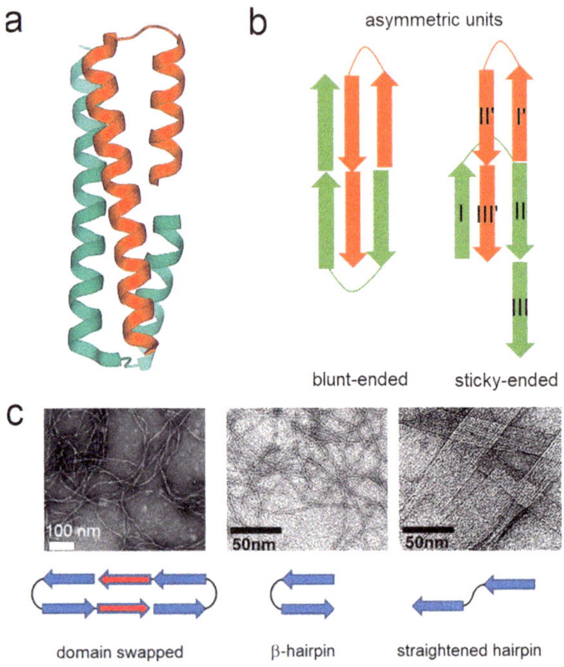

Figure 4.15 Domain-swapping designs. (a) A domain-swapped dimer (PDB entry 1G6U rendered by PyMol) forming an asymmetric three-helix bundle unit. (b) Schematic representations of the dimer in a blunt-ended format, highlighting the up–down–down topology of constituent helical domains (left), and in a sticky-ended format setting up the assembly of the helical hairpins (right). (c) Temperature-dependent assembly of a strand-swapping peptide into β-structured fibrils. Schematics (bottom) and electron micrographs (top) of different β-hairpin topologies. Red strands denote exchangeable domains. Adapted from ref. 166 and 168 with permission from American Chemical Society, Copyright 2005, 2008.

Similarly to other helical designs, the domain-swapping approach proves to be amenable to thermally switchable hydrogel designs. An advantage of this approach is that it is not limited to helix formation and can be used with other folding elements to the point of conformationally controlled assembly, such as β-structured fibrils induced by reversible β-hairpin folding.[166,167] Indeed, minimalist β-hairpins can be designed to readily adopt domain swapping. An asymmetric hairpin can be designed in the same manner in helical units, Figure 4.15b: one strand in the hairpin is made to extend such that it would stick to the same extension of another hairpin.[168] Thus, the resultant strand-swapping assembly unit contains an exchangeable β-strand domain coupled with a non-exchangeable β-hairpin domain, Figure 4.15c. At temperatures up to physiological, the peptide is predictably in a random coil conformation and remains soluble in water. As the temperature increases, the unit switches to a two-domain structure, in which the folded β-hairpin serves as a scaffold displaying the exchangeable strand. The latter strand becomes available to exchange with its counterpart in another unit copy with the formation of a bilayer of strand-swapped dimers. The dimer in turn propagates laterally into β-sheet-stacked fibrils, giving rise to a hydrogel.[168] By varying the length of the exchangeable strand, one can determine the mechanical properties of hydrogels, which can relate to the specific morphologies of assembled fibrils. For example, shorter domains forming non-twisted fibrils give hydrogels nearly two times more mechanically rigid than those composed of twisted ribbons produced by domains of doubled lengths.[167,168] The topology of β-hairpins appears to determine the nanoscale morphology of resultant materials in addition to their physicochemical properties. Short symmetric hairpins assemble into reversible fibre-rich hydrogels, but when strengthened the hairpins adopt an extended conformation, which converts into irreversible, laminated amyloid-like membrane sheets.[166]

Physical phase transitions find use in applications that require reversible materials. Self-assembling β-hairpins can be designed as injectable hydrogels undergoing multiphase transition. Once assembled, the gels can be introduced by shear flow using a syringe into the lumen of damaged blood vessels. The gel transits from a solid to a viscous gel in the process, but rapidly recovers in the lumen, where it distends the vessel, and in the space between two vessels, where it acts as a glue between the vessel ends. Suturing can then be performed through the gel, and upon completion light is used to trigger the gel–sol transition, disrupting the β-structured networks of the hydrogel, which degrades the gel and allows blood flow through the sutured vessel, Figure 4.16.

Biomimetic strategies rely on finding an exploitable synergy between positive designs, *i.e.* towards the desired structure, form or functions, and negative designs, *i.e.* away from alternatives. Therefore, structural originators for a desired biodesign do not necessarily support the function for which the design is meant. A notable example originates from a series of studies to elucidate the minimalist unit of amyloid fibril formation. The unit or a "core" sequence that is capable of nucleating amyloidogenesis was identified as a

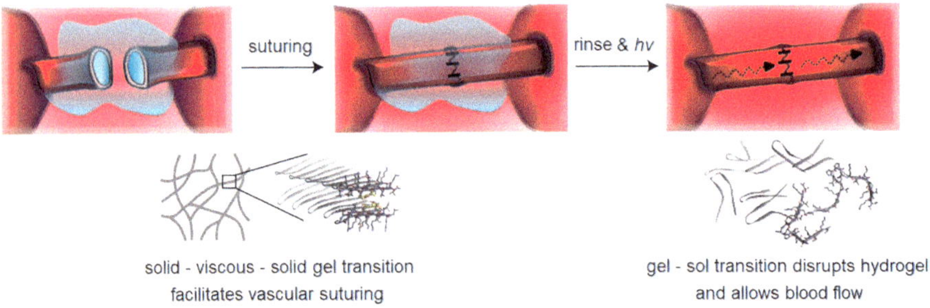

Figure 4.16 A multiphase transitioning peptide hydrogel for vascular suturing. Sol–gel phase transition triggers the formation of a solid-like hydrogel. Shear-thin injection into a damaged vessel makes it viscous. The gel then recovers into a solid hydrogel, facilitating suturing. Ultraviolet photolysis triggers the final gel–sol transition to allow blood flow. Adapted from ref. 169 with permission from Springer Nature, Copyright 2015.

dipeptide – phenylalanylphenylalanine (FF). The peptide was demonstrated experimentally to self-assemble into fibrillar nanotubes,[170] which confirmed earlier predictions that hydrophobic dipeptides, including the FF motif, are capable of forming hollow nanofibre structures.[171,172] It was also observed that hydrophobic dipeptides derivatized with an *N*-(fluorenyl-9-methoxycarbonyl) (Fmoc) group become efficient hydrogelators, which could be used as functionalized scaffolds in 3D cell culture and for related applications.[173–175] A general model for the assembly of short hydrophobic peptides has also been proposed. The peptides assemble into needle-like fibrils *via* the anti-parallel arrangement of hydrogen-bonded β-sheets and π–π stacking interactions of aromatic phenylalanine rings. For the Fmoc version of FF, Fmoc groups zip up into π-stacked alternate pairs with interleaved phenyl rings. With a slight twist of β-pleated structures, self-associating β-sheets rotate around each other, shaping a cylindrical structure.[176] The model also supports the notion that the morphology of nanostructures is defined by subtle differences in short peptide motifs, which can be explored to devise libraries of nanoscale morphologies and gelators with tuneable gelation properties.[177,178]

The pace with which the area of fibrous materials is progressing makes it practically impossible to cover all reported developments. Nonetheless, however, successful fibrillar systems, networks and gels provide only a partial answer to the design of artificial matrices. Questions remain as to the exact biology that needs to be built into a mimetic matrix to operate in synergy with specific morphological features designed as cues to instruct specific cell responses and aid in accommodating changes in dynamic cellular environments. Over recent decades, the term "regenerative medicine" has been applied liberally to embrace tissue engineering approaches aimed at accelerating natural healing processes.[179] The central paradigm of these approaches is therefore the use of ECM analogues in the native 3D cellular environment.

In technical terms, fibrous supramolecular scaffolds are designed to promote the seeding of tissue-specific cells that become spatially confined in 3D to develop further into living tissues. The rationale for most of these designs is therefore relatively straightforward – to grow in 3D, cells need to be embedded in a structure that mimics the ECM, which is ideally a 3D nanostructured fibrillar protein scaffold. Naturally derived fibrillar materials tested as ECM substituents have persistent limitations, including handling problems, morphological incongruity, technical difficulties concerned with engineering specific properties, potential immunogenicity and pathogen and disease transmission, which eventually prompted the design of artificial, bottom-up systems as a sound alternative.

Tissue engineering approaches imply the use of a scaffold to support the growth and proliferation of tissue-specific cells *ex vivo*. Such a cell-based definition is distinguished from guided tissue regeneration in which designed scaffolds direct and maintain the regeneration of a tissue only from cells residing in the site of transplantation.[180] These differences are evolving, giving rise to hybrid approaches and also complementary strategies of designing bioartificial tissues.

4.4 Non-matrix Mimetics of the ECM

The construction of bioartificial tissues aims at devising an appropriate replacement for a tissue through the use of biomaterial scaffolds that either are shaped like the damaged or missing tissue or organ (*ex vivo*) or, once deposited into the defect site, are able to adapt in a desired tissue environment (*in vivo*). The *ex vivo* approaches represent the concept of "spare body parts", which lends itself to a practice of replacing injured or lost organs. In this approach, cells are expanded *in vitro* and seeded onto the scaffold often of a preformed shape to mature into a tissue or organ prior to implantation into the patient.

The *in vivo* approaches are utilized to support continuous restoration of self-renewing tissues and rely on the use of injectable and responsive materials that by filling the injury site are arranged into a 3D supporting matrix to encourage the surrounding cells to develop into a living tissue *in situ*, Figure 4.17.[180] Injecting scaffolds laden with stem cells accelerates tissue restoration. This is also where self-healing properties of the scaffolds prove essential to respond to cell-instructed remodelling. In both of these cases, cell-supporting scaffolds are meant to mimic the ECM, and both strategies can adapt to any cell or tissue type, which has been demonstrated by successful solutions to the *ex vivo* constructions of cardiac valve substitutes, corneal transplants and urinary bladders and *in vivo* nerve, bone and cartilage regeneration.[179–184]

At the cellular level, ECM analogues are required to be differential, supporting a range of biological properties being able to promote cell adhesion and migration, sequester and release biological signals, and the like. However, there are other equally important criteria set by the use of the analogues

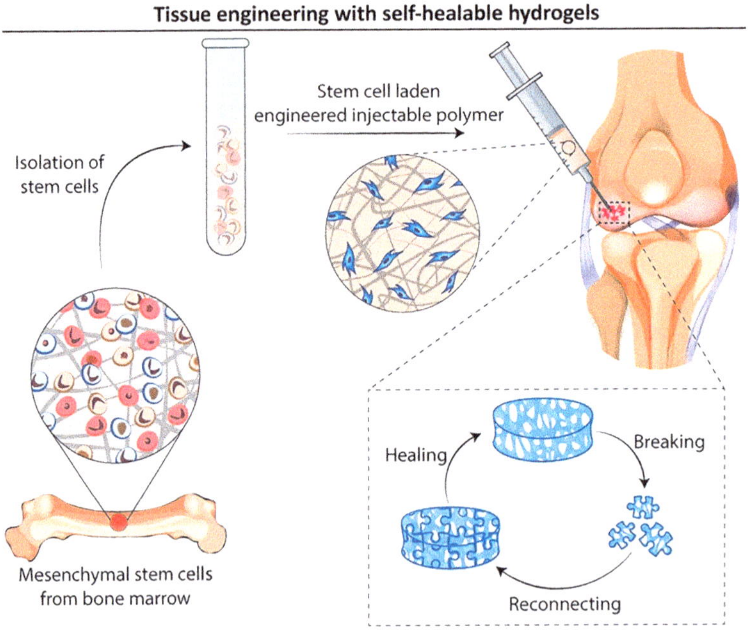

Figure 4.17 *In vivo* tissue engineering approach. Tissue engineering hydrogels designed to heal on command and incorporate mechanical and physicochemical properties that are native to tissues. If made injectable, the hydrogels can deliver stem cells to the injury site where the hydrogels can undergo remodelling in response to the cell development. Reproduced from ref. 180, https://doi.org/10.1002/advs.201801664, under the terms of the CC BY 4.0 license https://creativecommons.org/licenses/by/4.0/.

in clinical practice. Poor cell diffusion, risk of infection or disease transmission, mechanical and physical incompatibility of preformed scaffolds, immunorejection of the implanted material and invasive surgery are typical shortcomings leading to treatment complications and increased costs.

Living or "smart" materials may provide a saving solution, Figure 4.17. These are cell-supporting scaffolds assembled directly in a tissue from precursors, which allows them to integrate fully within an injury site or a tissue lesion. This approach entails the co-assembly of the scaffold with cells, which become homogeneously distributed within a hosting tissue cavity regardless of its irregularities of shape or size. Thus, a reciprocal nano-to microscale environment, in which the scaffold responds to cells and cells respond to the scaffold, is created. This makes invasive surgery, which otherwise is necessary for *ex vivo* approaches, practically excessive, Figure 4.17.[185] Based on these differences, open and closed artificial tissues can be defined.[186]

Open tissues are those that become vascularized by the host tissue and require temporary scaffolds to provide initial support for cell growth, and these scaffolds are subsequently replaced with the native ECM produced by the surrounding cells. The time periods to allow this can vary from several days to weeks, depending on the tissue type and application. Fibrous matrices and reversible hydrogels can be viewed as precursors for open systems.

Closed tissues arise from the slow diffusion of cells encapsulated into a scaffold that is resistant to degradation. Scaffolds used for closed tissues are typically porous microcapsules or membranes of $<500\ \mu m$ diameter and $<5\ \mu m$ thickness. These are hydrogels, *e.g.* alginates, or copolymeric shells, *e.g.* acrylonitriles and methylacrylates. A typical microcapsule design would comprise an inner cell-holding endoskeleton, *i.e.* matrix, and an outer exoskeleton, *i.e.* matrix-surrounding shell, with a charged biopolymer making up the inner layer, which constitutes a matrix component, and an oppositely charged synthetic polyelectrolyte shaping the outer layer of the capsule. The combination of the two gives a macroporous membrane. The porosity renders such a microcapsule permeable to nutrients for normal metabolic functioning of cells and to toxins released by the cells.

Both open and closed designs have been shown to give comparably efficient extracellular systems, which share many similarities in supporting normal cell development. Having both approaches available and developed gives additional flexibility in selecting appropriate strategies for regenerative medicine. However, in some areas the preferential use of one type over the other is justified.

For example, foreign xenogeneic or allogeneic cell materials, which are common for tissue engineering applications, can elicit immunorejection and should be expected to be better used in combination with closed constructions. The nature of closed systems permits immunoprotection, which is not characteristic of open scaffolds. On the other hand, open bioartificial tissues make better use of stem cells that need only a short-term boost for differentiation, Figure 4.17.

Open bioartificial tissues have attracted increasing attention largely owing to their compatibility with stem cells, in particular embryonic stem cells, which present an ideal source for transplantable cells. Stem cells can be expanded and maintained *in vitro* while retaining their ability for directed differentiation into any cell type in the body.[187–189]

Embryonic stem cells have been shown to promote the repair of ischaemic tissues, induce the production of dopamine-generating cells for Parkinson's disease, and promote the conversion to progenitor cells that were able to region-specifically differentiate inside brain into neurons and astrocytes, or to cardiomyocytes enhancing cardiac repair after myocardial infarction.[188–192] Open scaffolds can be most effective in cases where the implantation of other cells is to be avoided.

The combination of embryonic stem cells and open scaffolds is particularly beneficial for engineering tissues dependent on enhanced vascularization, such as muscle, liver and spleen. However, the use of human embryonic stem cells in regenerative therapies continues to stimulate ethical debates as human embryos are their only source.[187,192] In principle, the cells can be replaced with human adult stem cells, many of which can be cultured *in vitro*. Cell transplants from umbilical cord blood or bone marrow have already been found to cure certain cancers, aid in haematopoietic reconstitution after local tissue removal or destruction and support the construction of heart valves. However, adult stem cells are not devoid of drawbacks.[191] Not all tissues can be harvested, and those that can are subject to depletion over time. Emerging reports on the developmental plasticity of adult stem cells suggest that a single autogenic type could serve as a source for any differentiated cell type. However, the use of stem cells remains a matter of compromise. At the same time, increasing efforts are being focused on issues that are a prerequisite for constructing technically viable bioartificial tissues: notwithstanding differences in cells – stem, differentiated or immortalized – their use for tissue repair requires complex 3D extracellular environments.[193,194] Three-dimensional nanostructured matrices provide extracellular environments supporting physiological conditions of native cell surroundings.[195–197] The importance of 3D models is compounded by the fact that the outcomes in 3D models may differ from those obtained on 2D surfaces.[198] Cell migration in 2D is restricted to adhesion and translocation, whereas 3D environments allow one to assess the remodelling of the matrix in response to changes in the behaviour of migrating cells and also the mechanics of growing tissues.[199] Consequently, the use of 3D matrices helps reveal how matrix–cell–matrix signalling and interactions define tissue phenotypes, which cannot be unambiguously assessed on 2D substrates.

Contraction forces constitute the main factors driving the 3D organization of a tissue. These can be viewed as variational changes in the physical properties of the matrix occurring in response to cell spreading. The extent to which these occur is defined by spatial constraints, which are introduced by the matrix. Typically, ECM mimetics are designed to provide tensile strength to support cell attachment and proliferation, and also to resist mechanical stress wielded by cells. This may lead to a heterogeneous distribution of cells within a given matrix. Some matrix designs can accommodate the stress by exhibiting good self-healing properties, but most matrices respond solely with increases in density. In native tissues, this is compensated by the secretion of the ECM components and does not cause dramatic changes in the matrix volume. For example, density increases of collagen or fibrin fibres occur at the expense of increased protein concentrations and are proximal to surrounding cells, which allows the matrix to sustain mechanical tension to be controlled locally, *i.e.* by restraining or aligning fibres.[198] By contrast, in synthetic matrices such changes lead to decreases in the matrix volume, hampering the assembly of cells into a desired tissue, which gives rise to morphologically undefined materials with poor differentiation profiles.

The stumbling point here is the lack of control over the matrix–cell assembly and relationships between the matrix structure and tissue mechanics, which are defined at the nano-to microscale. The development of synthetic matrices is one way to address this. However, a paradigm shift is observed in much research focused on devising technologies promoting the cell-induced production of the ECM,[200] with two mainstream approaches being 3D micropatterning and *in vivo* bioreactors.

Micropatterning is a collective term for miniaturization techniques, which can be based on top-down and bottom-up, 2D and 3D strategies.[201,202] This implies the use of temporary matrices that, as described earlier, must fully integrate into the host tissue. Temporary matrices set up tissue architecture in 3D, induce the secretion and assembly of the native ECM and subsequently deteriorate, leaving the maintenance of the new tissue to the ECM produced by the cells. Although this is an idealized view, any temporary matrix would pursue solutions to technical challenges, which are identical with those posed by the structural problem of an artificial matrix and would implement the found solution in the positional context of a living tissue.

As an exemplar, 3D matrices can be micropatterned to initiate and control the geometry of mammary epithelial tubules.[203] Epithelial cells embedded in a gel collagenous matrix generated around a patterned elastomeric stamp tend to form hollow tubules conforming to the size and shape of the matrix cavities. When branching micropatterns are introduced, multicellular branches form that extend into the surrounding collagen. The branches initiate only from the ends and not the sides of the tubules, which is characteristic of the dichotomous branching of mammary end buds *in vivo*. The tubules are thus polarized, which aids in the directional cell invasion of the surrounding tissues.

Although neutralizing the activity of factors necessary for branching morphogenesis such as matrix metalloproteinases or mesenchymal morphogens proved to prevent the invasion, branching was retained. This indicates that branching micropatterns instruct the cells to branch according to their position within the tubule. Furthermore, increases in the length of the tubules had no effect on the polarity of branching, even though increased branching density was observed. Curved tubules branched from the convex side of the curve, whereas bifurcated tubules branched from distal positions. This study exemplifies the role of a 3D matrix in setting up and controlling the geometry and positioning of epithelial tissue development, which in physiological terms relates to the transformation of epithelial tubule rudiments to mature ductal tree. Branching morphogenesis is universal across many cell types, and provides a demonstrable model to study cell behaviour in 2D and 3D. For example, the flattened appearance of fibroblasts on planar surfaces tends to be markedly different from that of fibroblasts in connective tissues. In contrast to the lamellar shape of a 2D fibroblast, dendritic networks with extended and slender branches are characteristic for cells observed *in vivo*. This difference in appearance is attributed to how cells respond to nanometre features on the surface of their substrates, and to local surface stiffness, which in 2D permits increased cell spreading and focal adhesion, but is not typical for 3D environments, in which cells are under isometric tension only during fibrotic conditions such as wound healing.[204]

A stress-accommodating, floating 3D matrix with high cell-embedding capacity (typically $>10^6$ mL^{-1}) was used to help fibroblasts adopt a more native morphology. The cell embedded in the matrix and produced highly dynamic dendritic networks the extensions of which appeared to project and retract.[205] In contrast, no dendritic projections were observed in 2D surfaces. Cell analysis in restrained as opposed to floating matrices at low (10^5 mL^{-1}) and high cell density (10^6 mL^{-1}) revealed that matrix contraction was apparent only at high cell density, in which case dendritic networks appeared to be initial fibroblast morphology that over time converted to bipolar stellate structures. This observation is common for micropatterned and structurally restrained scaffolds but not for compliant and stress-responsive matrices.

Using the analogy of neurons the dendritic morphology of which is stabilized by actin and microtubules, the dependence of the structure of the fibroblast dendritic networks on the cytoskeleton components can also be assessed. It was found that cell extensions contained a core of microtubules with actin localized at the tips. The phenomenon of the formation and continued stabilization of dendritic networks mediated by both actin microfilaments and microtubules was not observed in 2D and could be mimicked only using 3D matrices. In particular, the concentration of actin at the tips of the microtubule core can serve as another example of positional context in directing branching morphogenesis.

Positional setting can be extended to multiple micropatterning to permit interfacing multiple, but not mixing, 3D matrices for tissue modelling.[206] With native tissues composed of multiple ECM types, the use of patterned native ECM phases can help counter cellular contractile forces. To achieve this, a matrix is required that is capable of incorporating multiple microscale patterns and also interfacing individual phases in the matrix. Collagen fibres in a preformed ECM may act as nucleation sites for the assembly of collagen fibres in an adjacent ECM gel. Adjacent matrices thus are to be used for the integration of adjacent 3D ECM phases, that is, matrices at the phase boundary would act as templates for the assembly of new matrices within the bulk ECM.

A proposed construct meeting these requirements used collagen fibres, which constituted patterned phases, flowing into microfluidic alginate hydrogels, which were used as gelling or bulk phases. It this design, the gelation of the patterned phase occurred at the expense of collagen fibres crossing from the bulk phase boundary. In contrast, when a collagen-free alginate was used as a bulk phase, patterned fibres formed only in the flowing ECM solution and were not supplied by the bulk boundary. This supports the feeding model of new collagen fibres that tended to be assembled at the boundary by pre-existing collagen, which was accumulated at the interface.

The model was used to probe whether collagen fibre assembly could link natural ECM phases together to resist cell contraction. Fibroblasts and endothelial cells were seeded into alginate as a bulk phase and channel-patterned collagen I matrices as a patterned phase, respectively. Fibroblasts were uniformly distributed in the bulk phases, whereas the distribution of endothelial cells was spatially restricted to the micropatterned phase. In collagen-free bulk phases, the patterned collagen was found to be detached from the bulk interface and significantly contracted after a few hours of culturing, while no further migration of endothelial cells was observed after the contraction. In marked contrast, continued migration and primary proliferation (upon addition of angiogenic growth factors) were observed in the collagen-loaded bulk phases, where the phase interfaces remained integrated over the course of several weeks.

The method represents an excellent interfacial model of diffusively permeable phases that can be used for studying cell migration and proliferation in response to designed external stimuli. Designs of this type hold promise for developing 3D multiple ECM phases, the regularity and reconstruction of which require the flexibility of more than one assembly mode. The latter is probably the reason why most micropatterning techniques still employ native ECM assemblies – collagen or fibrin matrices. A compelling case of artificial self-assembly systems based on helical, β-structured peptides or peptidic amphiphiles to 3D matrix micropatterning has yet to be demonstrated.

Alternative materials may derive from natural or synthetic polymers. A 3D porous scaffold would be an appropriate 3D architecture to pattern. The main attraction of polymer-based materials is the relative ease with which their porosity and pore sizes can be determined. The fabrication of polymer-based scaffolds relies largely on solidification, cross-linking and morphological preshaping. Therefore, porosity becomes the prime property that puts this type into the category of open and temporary matrices.

Porosity is necessary to support cell migration and proliferation. The freedom of migration for cells is more substantial in open systems than that which can be provided by closed systems, which may only support permeability sufficient for the uptake of nutrients and removal of metabolic waste. Porosity is also necessary for the production of the native ECM and can be tailored to respond to changing proliferation rates of the seeded cells. Providing reciprocal mechanotransduction between scaffolds and cells is an intrinsic characteristic of dynamically equilibrated self-assembled matrices but is the main hurdle for polymer-based scaffolds. Porous polymer scaffolds can be ideal systems for engineering solid tissues, *e.g.* cartilage or bone. However, this is largely attributable to their excellent grafting properties rather than to efficient nano-to microscale patterning, which prompted the emergence of an equally effective approach for engineering bioartificial tissues – *in vivo* bioreactors.

Arguably, the drawbacks of grafting strategies are what stimulated the development of *in vivo* bioreactors. As an exemplar, traditional bone grafting is reliant on auto- and allografts utilized to repair bone losses, spinal fusions and long-bone fractures. Although bone grafts have substantially advanced bone engineering, their use is associated with complications, which mostly derive from the imperfections of graft materials leading to tissue devascularization, risks of infections or morbidity at donor sites.[207,208]

Graft strategies have evolved into the concept of an ideal designer graft material that can be formulated as the sum of three interrelated elements: cell-conductive matrix, cell-conductive proteins and committed cells; all three being pooled together make up a simplified version of an *in vivo* bioreactor.

Such a reactor can be made as a sealed porous or conductive matrix assembled with a vascular pedicle and impenetrable silicone shield. The bioreactor can then be introduced into a living tissue and coated with growth factors, *i.e.* conductive proteins, with the recruitment of osteogenic cells, thus capitalizing on the neovascularization of the scaffold acting as a conduit. The bioreactor is not a closed bioartificial tissue as it may seem but a sealed open tissue since its main purpose, to emulate bone formation, occurs in an isolated environment and the surrounding tissue serves as a supply of cells invading the reactor.[209] Indeed, the cells not only successfully invade the scaffold but also spread and proliferate within, with the subsequent production of the native ECM – the main characteristic of a living open tissue. The design represents a very good example of how vascularization can be initiated and maintained within a short period of time, but which is sufficient for generating a living tissue. It also advocates an *in vivo* approach to engineer bone as *in vitro* techniques fail to spatially relate forming blood vessels to the architectural integrity of bone. With many attempts at engineering functional bone *in vitro* stumbling at the pivotal reliance of bone tissue on vascularization, it is clear that the development of different cell types into one tissue architecture has to present an alternative process. The search for solutions to this challenge stimulated the development of porous polymeric materials in combination with bone stem cells and morphogenetic proteins. This strategy provides a straight route to the interplay between scaffolds, cells and growth factors and hence to creating environments that promote functional tissue regeneration.[210,211]

Interestingly, however, this transplantation formula, scaffold-cell growth factors, in bone engineering can be avoided. Large volumes of bone can be produced without the need for cell transplantation and the administration of growth factor. The core of this particular strategy is not an introduction or injection but the creation of an *in vivo* bioreactor within the body.[212] The bioreactor in this case is a space, which is deliberately formed between the surface of a long bone and its membrane, that is rich in stem cells. Thus, the required cell population and all necessary growth signals are prelocalized or derived locally. The idea of constructing a bioreactor by simply generating a space of chosen geometry within the body is an attractive alternative to exogenously introduced bioreactors. However, the technical realization of the concept proved to require a comparable external intervention. First, a hydraulic elevation procedure was required to generate the space.[213] Second, once made, the space was filled with a cell-supporting matrix, alginate gel, followed by its cross-linking, which makes the proposed bioreactor at least a two-component system that cannot function without an injectable 3D matrix, Figure 4.17.

Possibly owing to their intrinsic conceptual limitations or relatively short history, *in vivo* bioreactor designs have yet to progress beyond laboratory development. Distinctively supportive of the concept in this regard are investigations targeting other issues in cell and tissue biology; for example, in providing 3D biomimetic models that can find use in establishing the effects of microenvironmental conditions and nanoscale patterning on tumour malignancy *in vivo* as compared with those established *in vitro*.[214]

4.5 Outlook

Assessing the progress of regenerative medicine through the prism of *in vivo* bioreactors emphasizes the need for the design of smart ECM mimetics, which can operate at the nano-to micrometre length scales to reciprocate cell responses. With their complexity and promise, *in vivo* bioreactors constitute an optimal strategy to develop. With their relatively moderate technological advancement, they reinstate the midway point between the chemical mimicry of the ECM and the clinical implementation of regenerative therapies. The key for success appears to be that the development of more efficient bioreactors converts the technical challenges facing artificially guided regeneration into an all-encompassing problem of creating a suitable live ECM mimetic. The task of constructing such a mimetic, which can be controlled as a responsive extracellular niche, which can control tissue repair, is multifold. Its structure, from the nano-to macroscale, its morphological versatility, compatibility with cells and mass transport, and its ability to remodel, self-heal and maintain vascularization are among just a few persistent barriers that remain between the potential and the reality of rationally induced regeneration, between "today's Frankenstein and tomorrow's benefactor".[186] However, an important question remains regarding the need for or even a viable possibility of extracellular matrices to provide a solution to regeneration, or perhaps a more effective solution lies in achieving better control over biology regardless of the materials used. Future biological designs with properties defined at the nanoscale hold promise in addressing all these questions as ultimately regeneration capabilities, dormant or active, are designed within the space of a single cell.

References

1. R. J. Goss, *Principles of Regeneration*, Academic Press, New York, 1969.
2. R. J. Goss, *The Natural History (and Mystery) of Regeneration*, ed. Dinsmore, C. E., Cambridge University Press, Cambridge, 1991.
3. E. M. Tanaka, Regeneration: If They Can Do It, Why Can't We?, *Cell*, 2003, **113**, 559–562.
4. O. J. Reichman, Evolution of Regeneration Capabilities, *Am. Nat.*, 1984, **123**, 752–763.
5. R. Langer and D. A. Tirrell, Designing materials for biology and medicine, *Nature*, 2004, **428**, 487–492.
6. A. S. Alvarado and P. A. Tsonis, Bridging the regeneration gap: genetic insights from diverse animal models, *Nat. Rev. Genet.*, 2006, **7**, 873–884.

7. P. A. Tsonis, Regenerative biology: the emerging field of tissue repair and restoration, *Differentiation*, 2002, **70**, 397–409.
8. P. A. Tsonis, Regeneration in Vertebrates, *Dev. Biol.*, 2000, **221**, 273–284.
9. G. C. Gurtner, S. Werner, Y. Barrandon and M. T. Longaker, Wound repair and regeneration, *Nature*, 2008, **453**, 314–321.
10. D. L. Stocum, Tissue restoration through regenerative biology and medicine, *Adv. Anat., Embryol. Cell Biol.*, 2004, **176**, 1–101.
11. J. R. Erickson and K. Echeverri, Learning from regeneration research organisms: The circuitous road to scar free wound healing, *Dev. Biol.*, 2018, **433**, 144–154.
12. S. Levenberg, R. Langer and P. S. Gerald, Advances in Tissue Engineering, *Curr. Top. Dev. Biol.*, 2004, **61**, 113–134.
13. Z. Lansky, Y. Mutsafi, L. Houben, T. Ilani, G. Armony, S. G. Wolf and D. Fass, 3D mapping of native extracellular matrix reveals cellular responses to the microenvironment, *J. Struct. Biol.: X*, 2019, **1**, 100002.
14. H. K. Kleinman, D. Philp and M. P. Hoffman, Role of the extracellular matrix in morphogenesis, *Curr. Opin. Biotechnol.*, 2003, **14**, 526–532.
15. A. S. Mao and D. J. Mooney, Regenerative medicine: Current therapies and future directions, *Proc. Natl. Acad. Sci. U. S. A.*, 2015, **112**, 14452–14459.
16. *Extracellular Matrix Assembly and Structure*, ed. Yurchenco, P. D., Birk, D. E. and Mecham, R. P., Academic, San Diego, 1994.
17. G. N. Ramachandran and G. Kartha, Structure of Collagen, *Nature*, 1954, **174**, 269–270.
18. G. N. Ramachandran and G. Kartha, Structure of Collagen, *Nature*, 1955, **176**, 593–595.
19. A. Rich and F. H. C. Crick, The Structure of Collagen, *Nature*, 1955, **176**, 915–916.
20. S. Ricard-Blum, The collagen family, *Cold Spring Harbor Perspect. Biol.*, 2011, **3**, a004978.
21. M. van der Rest and R. Garrone, Collagen family of proteins, *FASEB J.*, 1991, **5**, 2814–2823.
22. D. F. Holmes, H. K. Graham, J. A. Trotter and K. E. Kadler, STEM/TEM studies of collagen fibril assembly, *Micron*, 2001, **32**, 273–285.
23. D. F. Holmes, *et al.*, Corneal collagen fibril structure in three dimensions: Structural insights into fibril assembly, mechanical properties, and tissue organization, *Proc. Natl. Acad. Sci. U. S. A.*, 2001, **98**, 7307–7312.
24. L. M. Shaw and B. R. Olsen, FACIT collagens: diverse molecular bridges in extracellular matrices, *Trends Biochem. Sci.*, 1991, **16**, 191–194.
25. C. Wang, *et al.*, The third activity for lysyl hydroxylase 3: galactosylation of hydroxylysyl residues in collagens *in vitro*, *Matrix Biol.*, 2002, **21**, 559–566.
26. J. Baum and B. Brodsky, Folding of peptide models of collagen and misfolding in disease, *Curr. Opin. Struct. Biol.*, 1999, **9**, 122–128.

27. T. Starborg, Y. Lu, R. S. Meadows, K. E. Kadler and D. F. Holmes, Electron microscopy in cell-matrix research, *Methods*, 2008, **45**, 53–64.
28. H. N. Su, L. Y. Ran, Z. H. Chen, Q. L. Qin, M. Shi, X. Y. Song, X. L. Chen, Y. Z. Zhang and B. B. Xie, The ultrastructure of type I collagen at nanoscale: large or small D-spacing distribution?, *Nanoscale*, 2014, **6**, 8134–8139.
29. S. Köster, D. A. Weitz, R. D. Goldman, U. Aebi and H. Herrmann, Intermediate filament mechanics *in vitro* and in the cell: from coiled coils to filaments, fibers and networks, *Curr. Opin. Cell Biol.*, 2015, **32**, 82–91.
30. M. Pieters and A. S. Wolberg, Fibrinogen and fibrin: An illustrated review, *Res. Pract. Thromb. Haemostasis*, 2019, **3**, 161–172.
31. A. M. Hindeleh and R. Hosemann, Paracrystals representing the physical state of matter, *J. Phys. C: Solid State Phys.*, 1988, **21**, 4155–4170.
32. M. K. Gardner, A. J. Hunt, H. V. Goodson and D. J. Odde, Microtubule assembly dynamics: new insights at the nanoscale, *Curr. Opin. Cell Biol.*, 2008, **20**, 64–70.
33. H. K. Graham, D. F. Holmes, R. B. Watson and K. E. Kadler, Identification of collagen fibril fusion during vertebrate tendon morphogenesis. The process relies on unipolar fibrils and is regulated by collagen-proteoglycan interaction, *J. Mol. Biol.*, 2000, **295**, 891–902.
34. D. F. Holmes, M. P. Lowe and J. A. Chapman, Vertebrate (chick) collagen fibrils formed in Vivo can exhibit a reversal in molecular polarity, *J. Mol. Biol.*, 1994, **235**, 80–83.
35. K. E. Kadler, Y. Hojima and D. J. Prockop, Collagen fibrils *in vitro* grow from pointed tips in the C- to N-terminal direction, *Biochem. J.*, 1990, **268**, 339–343.
36. J. W. Weisel, The mechanical properties of fibrin for basic scientists and clinicians, *Biophys. Chem.*, 2004, **112**, 267–276.
37. J. W. Weisel, C. Nagaswami and L. Makowski, Twisting of fibrin fibers limits their radial growth, *Proc. Natl. Acad. Sci. U. S. A.*, 1987, **84**, 8991–8995.
38. E. A. Ryan, L. F. Mockros, J. W. Weisel and L. Lorand, Structural Origins of Fibrin Clot Rheology, *Biophys. J.*, 1999, **77**, 2813–2826.
39. J. Halper and M. Kjaer, Basic components of connective tissues and extracellular matrix: elastin, fibrillin, fibulins, fibrinogen, fibronectin, laminin, tenascins and thrombospondins, *Adv. Exp. Med. Biol.*, 2014, **802**, 31–47.
40. F. Ramirez, L. Y. Sakai, H. C. Dietz and D. B. Rifkin, Fibrillin microfibrils: multipurpose extracellular networks in organismal physiology, *Physiol. Genomics*, 2004, **19**, 151–154.
41. L. Y. Sakai, D. R. Keene, M. Renard and J. De Backer, FBN1: The disease-causing gene for Marfan syndrome and other genetic disorders, *Gene*, 2016, **591**, 279–291.
42. K. Wolf, Y. I. Wu, Y. Liu, J. Geiger, E. Tam, C. Overall, M. S. Stack and P. Friedl, Multi-step pericellular proteolysis controls the transition from individual to collective cancer cell invasion, *Nat. Cell Biol.*, 2007, **9**, 893–904.

43. N. Naik, J. Caves, E. L. Chaikof and M. G. Allen, Generation of spatially aligned collagen fiber networks through microtransfer molding, *Adv. Healthcare Mater.*, 2014, **3**, 367–374.
44. V. A. Kumar, J. M. Caves, C. A. Haller, E. Dai, L. Liu, S. Grainger and E. L. Chaikof, Acellular vascular grafts generated from collagen and elastin analogs, *Acta Biomater.*, 2013, **9**, 8067–8074.
45. Z. Zhou, D. Qian and M. Minary-Jolandan, Molecular Mechanism of Polarization and Piezoelectric Effect in Super-Twisted Collagen, *ACS Biomater. Sci. Eng.*, 2016, **2**, 929–936.
46. H. K. Ravi, F. Simona, J. Hulliger and M. Cascella, Molecular origin of piezo- and pyroelectric properties in collagen investigated by molecular dynamics simulations, *J. Phys. Chem. B*, 2012, **116**, 1901–1907.
47. M. Takeuchi, M. Sekino, N. Iriguchi and S. Ueno, Dependence of the Spin-Spin Relaxation Time of Water in Collagen Gels on Collagen Fiber Directions, *Magn. Reson. Med. Sci.*, 2004, **3**, 153–157.
48. J. A. Fallas, L. E. O'Leary and J. D. Hartgerink, Synthetic collagen mimics: self-assembly of homotrimers, heterotrimers and higher order structures, *Chem. Soc. Rev.*, 2010, **39**, 3510–3527.
49. F. W. Kotch and R. T. Raines, Self-assembly of synthetic collagen triple helices, *Proc. Natl. Acad. Sci. U. S. A.*, 2006, **103**, 3028–3033.
50. B. K. Mann, Biologic gels in tissue engineering, *Clin. Plast. Surg.*, 2003, **30**, 601–609.
51. V. Bhagat and M. L. Becker, Degradable Adhesives for Surgery and Tissue Engineering, *Biomacromolecules*, 2017, **18**, 3009–3039.
52. O. Chaudhuri, J. Cooper-White, P. A. Janmey, D. J. Mooney and V. B. Shenoy, Effects of extracellular matrix viscoelasticity on cellular behaviour, *Nature*, 2020, **584**, 535–546.
53. K. Y. Lee and D. J. Mooney, Alginate: properties and biomedical applications, *Prog. Polym. Sci.*, 2012, **37**, 106–126.
54. B. Thu, *et al.*, Alginate gels – Some structure-function correlations relevant to their use as immobilization matrix for cells, *Prog. Biotechnol.*, 1996, **11**, 19–30.
55. P. Gacesa, Alginates, *Carbohydr. Polym.*, 1988, **8**, 161–182.
56. F. Brandl, F. Sommer and A. Goepferich, Rational design of hydrogels for tissue engineering: Impact of physical factors on cell behavior, *Biomaterials*, 2007, **28**, 134–146.
57. K. Y. Lee and D. J. Mooney, Hydrogels for Tissue Engineering, *Chem. Rev.*, 2001, **101**, 1869–1880.
58. K. H. Bouhadir, *et al.*, Degradation of Partially Oxidized Alginate and Its Potential Application for Tissue Engineering, *Biotechnol. Prog.*, 2001, **17**, 945–950.
59. H. J. Kong, D. Kaigler, K. Kim and D. J. Mooney, Controlling Rigidity and Degradation of Alginate Hydrogels *via* Molecular Weight Distribution, *Biomacromolecules*, 2004, **5**, 1720–1727.
60. J. L. Drury, T. Boontheekul and D. J. Mooney, Cellular cross-linking of peptide modified hydrogels, *J. Biomech. Eng.*, 2005, **127**, 220–228.

61. M. P. Lutolf and J. A. Hubbell, Synthetic biomaterials as instructive extracellular microenvironments for morphogenesis in tissue engineering, *Nat. Biotechnol.*, 2005, **23**, 47–55.
62. W. E. Hennink and C. F. van Nostrum, Novel crosslinking methods to design hydrogels, *Adv. Drug Delivery Rev.*, 2002, **54**, 13–36.
63. J. J. Rice, M. M. Martino, L. De Laporte, F. Tortelli, P. S. Briquez and J. A. Hubbell, Engineering the regenerative microenvironment with biomaterials, *Adv. Healthcare Mater.*, 2013, **2**, 57–71.
64. E. Nicol, Photopolymerized Porous Hydrogels, *Biomacromolecules*, 2021, **22**, 1325–1345.
65. K. T. Nguyen and J. L. West, Photopolymerizable hydrogels for tissue engineering applications, *Biomaterials*, 2002, **23**, 4307–4314.
66. M. Tortora, F. Cavalieri, E. Chiessi and G. Paradossi, Michael-type addition reactions for the in situ formation of poly(vinyl alcohol)-based hydrogels, *Biomacromolecules*, 2007, **8**, 209–214.
67. R. J. Christie, *et al.*, Photopolymerized hydrogel carriers for live vaccine ballistic delivery, *Vaccine*, 2006, **24**, 1462–1469.
68. D. J. Heath, P. Christian and M. Griffin, Involvement of tissue transglutaminase in the stabilisation of biomaterial/tissue interfaces important in medical devices, *Biomaterials*, 2002, **23**, 1519–1526.
69. M. P. Lutolf, G. P. Raeber, A. H. Zisch, N. Tirelli and J. A. Hubbell, Cell-responsive synthetic hydrogels, *Adv. Mater.*, 2003, **15**, 888–892.
70. M. P. Lutolf, *et al.*, Repair of bone defects using synthetic mimetics of collagenous extracellular matrices, *Nat. Biotechnol.*, 2003, **21**, 513–518.
71. G. Sorrentino, S. Rezakhani, E. Yildiz, S. Nuciforo, M. H. Heim, M. P. Lutolf and K. Schoonjans, Mechano-modulatory synthetic niches for liver organoid derivation, *Nat. Commun.*, 2020, **11**, 3416.
72. M. Hofer and M. P. Lutolf, Engineering organoids, *Nat. Rev. Mater.*, 2021, **19**, 1–19.
73. N. Gjorevski, N. Sachs, A. Manfrin, S. Giger, M. E. Bragina, P. Ordóñez-Morán, H. Clevers and M. P. Lutolf, Designer matrices for intestinal stem cell and organoid culture, *Nature*, 2016, **539**, 560–564.
74. N. Gjorevski and M. P. Lutolf, Synthesis and characterization of well-defined hydrogel matrices and their application to intestinal stem cell and organoid culture, *Nat. Protoc.*, 2017, **12**, 2263–2274.
75. J. Kopeček and J. Yang, Polymer nanomedicines, *Adv. Drug Delivery Rev.*, 2020, **156**, 40–64.
76. J. Kopeček and J. Yang, Smart self-assembled hybrid hydrogel biomaterials, *Angew. Chem., Int. Ed. Engl.*, 2012, **51**, 7396–7417.
77. W. A. Petka, J. L. Harden, K. P. McGrath, D. Wirtz and D. A. Tirrell, Reversible Hydrogels from Self-Assembling Artificial Proteins, *Science*, 1998, **281**, 389–392.
78. W. Shen, K. Zhang, J. A. Kornfield and D. A. Tirrell, Tuning the erosion rate of artificial protein hydrogels through control of network topology, *Nat. Mater.*, 2006, **5**, 153–158.

79. C. Wang, R. J. Stewart and J. Kopecek, Hybrid hydrogels assembled from synthetic polymers and coiled-coil protein domains, *Nature*, 1999, **397**, 417–420.
80. A. P. Nowak, *et al.*, Rapidly recovering hydrogel scaffolds from self-assembling diblock copolypeptide amphiphiles, *Nature*, 2002, **417**, 424–428.
81. E. Katchalski and M. Sela, Synthesis and chemical properties of poly-alpha-amino acids, *Adv. Protein Chem.*, 1958, **13**, 243–492.
82. T. J. Deming, Preparation and development of block copolypeptide vesicles and hydrogels for biological and medical applications, *Wiley Interdiscip. Rev.: Nanomed. Nanobiotechnol.*, 2014, **6**, 283–297.
83. W. J. Dearnaley, B. Schleupner, A. C. Varano, N. A. Alden, F. Gonzalez, M. A. Casasanta, B. E. Scharf, M. J. Dukes and D. F. Kelly, Liquid-Cell Electron Tomography of Biological Systems, *Nano Lett.*, 2019, **19**, 6734–6741.
84. K. Hammond, M. G. Ryadnov and B. W. Hoogenboom, Atomic force microscopy to elucidate how peptides disrupt membranes, *Biochim. Biophys. Acta, Biomembr.*, 2021, **1863**, 183447.
85. M. Shaw, A. Bella and M. G. Ryadnov, CREIM: Coffee Ring Effect Imaging Model for Monitoring Protein Self-Assembly in Situ, *J. Phys. Chem. Lett.*, 2017, **8**, 4846–4851.
86. A. Bella, M. Shaw, S. Ray and M. G. Ryadnov, Filming protein fibrillogenesis in real time, *Sci. Rep.*, 2014, **4**, 7529.
87. T. Koide, D. L. Homma, S. Asada and K. Kitagawa, Self-complementary peptides for the formation of collagen-like triple helical supramolecules, *Bioorg. Med. Chem. Lett.*, 2005, **15**, 5230–5233.
88. F. W. Kotch, I. A. Guzei and R. T. Raines, Stabilization of the Collagen Triple Helix by O-Methylation of Hydroxyproline Residues, *J. Am. Chem. Soc.*, 2008, **130**, 2952–2953.
89. S. Rele, *et al.*, D-Periodic Collagen-Mimetic Microfibers, *J. Am. Chem. Soc.*, 2007, **129**, 14780–14787.
90. L. E. O'Leary, J. A. Fallas, E. L. Bakota, M. K. Kang and J. D. Hartgerink, Multi-hierarchical self-assembly of a collagen mimetic peptide from triple helix to nanofibre and hydrogel, *Nat. Chem.*, 2011, **3**, 821–828.
91. D. R. Walker, S. A. H. Hulgan, C. M. Peterson, I. C. Li, K. J. Gonzalez and J. D. Hartgerink, Predicting the stability of homotrimeric and heterotrimeric collagen helices, *Nat. Chem.*, 2021, **13**, 260–269.
92. J. A. Fallas and J. D. Hartgerink, Computational design of self-assembling register-specific collagen heterotrimers, *Nat. Commun.*, 2012, **3**, 1087.
93. P. M. Gilbert, *et al.*, Substrate elasticity regulates skeletal muscle stem cell self-renewal in culture, *Science*, 2010, **329**, 1078–1081.
94. M. Guvendiren and J. A. Burdick, Engineering synthetic hydrogel microenvironments to instruct stem cells, *Curr. Opin. Biotechnol.*, 2013, **24**, 841–846.
95. J. Boekhoven and S. I. Stupp, 25th Anniversary Article: Supramolecular Materials for Regenerative Medicine, *Adv. Mater.*, 2014, **26**, 1642–1659.

96. Y. Li and S. M. Yu, Targeting and mimicking collagens *via* triple helical peptide assembly, *Curr. Opin. Chem. Biol.*, 2013, **17**, 968–975.
97. B. H. San, J. Hwang, S. Sampath, Y. Li, L. L. Bennink and S. M. Yu, Self-Assembled Water-Soluble Nanofibers Displaying Collagen Hybridizing Peptides, *J. Am. Chem. Soc.*, 2017, **139**, 16640–16649.
98. A. M. Hilderbrand, P. A. Taylor, F. Stanzione, M. LaRue, C. Guo, A. Jayaraman and A. M. Kloxin, Combining simulations and experiments for the molecular engineering of multifunctional collagen mimetic peptide-based materials, *Soft Matter*, 2021, **17**, 1985–1998.
99. E. De Santis and M. G. Ryadnov, Peptide self-assembly for nanomaterials: the old new kid on the block, *Chem. Soc. Rev.*, 2015, **44**, 8288–8300.
100. A. D. Merg, E. van Genderen, A. Bazrafshan, H. Su, X. Zuo, G. Touponse, T. B. Blum, K. Salaita, J. P. Abrahams and V. P. Conticello, Seeded Heteroepitaxial Growth of Crystallizable Collagen Triple Helices: Engineering Multifunctional Two-Dimensional Core-Shell Nanostructures, *J. Am. Chem. Soc.*, 2019, **141**, 20107–20117.
101. V. Conticello, S. Hughes and C. Modlin, Biomaterials Made from Coiled-Coil Peptides, *Subcell. Biochem.*, 2017, **82**, 575–600.
102. S. Zhang, T. Holmes, C. Lockshin and A. Rich, Spontaneous assembly of a self-complementary oligopeptide to form a stable macroscopic membrane, *Proc. Natl. Acad. Sci. U. S. A.*, 1993, **90**, 3334–3338.
103. S. Zhang and A. Rich, Direct conversion of an oligopeptide from a beta-sheet to an alpha-helix: A model for amyloid formation, *Proc. Natl. Acad. Sci. U. S. A.*, 1997, **94**, 23–28.
104. A. Aggeli, *et al.*, Responsive gels formed by the spontaneous self-assembly of peptides into polymeric beta-sheet tapes, *Nature*, 1997, **386**, 259–262.
105. M. J. Pandya, *et al.*, Sticky-End Assembly of a Designed Peptide Fiber Provides Insight into Protein Fibrillogenesis, *Biochemistry*, 2000, **39**, 8728–8734.
106. R. P. Nagarkar, R. A. Hule, D. J. Pochan and J. P. Schneider, Domain swapping in materials design, *Biopolymers*, 2010, **94**, 141–155.
107. N. L. Ogihara, *et al.*, Design of three-dimensional domain-swapped dimers and fibrous oligomers, *Proc. Natl. Acad. Sci. U. S. A.*, 2001, **98**, 1404–1409.
108. R. P. Nagarkar, R. A. Hule, D. J. Pochan and J. P. Schneider, De Novo Design of Strand-Swapped beta-Hairpin Hydrogels, *J. Am. Chem. Soc.*, 2008, **130**, 4466–4474.
109. S. Kojima, Y. Kuriki, T. Yoshida, K. Yazaki and K. Miura, Fibril formation by an amphipathic alpha-helix-forming polypeptide produced by gene engineering, *Proc. Jpn. Acad., Ser. B, Phys. Biol. Sci.*, 1997, **73**, 7–11.
110. H. Yokoi, T. Kinoshita and S. Zhang, Dynamic reassembly of peptide RADA16 nanofiber scaffold, *Proc. Natl. Acad. Sci. U. S. A.*, 2005, **102**, 8414–8419.
111. M. Altman, P. Lee, A. Rich and S. Zhang, Conformational behavior of ionic self-complementary peptides, *Protein Sci.*, 2000, **9**, 1095–1105.

112. S. Zhang, C. Lockshin, R. Cook and A. Rich, Unusually stable beta-sheet formation in an ionic self-complementary oligopeptide, *Biopolymers*, 1994, **34**, 663–672.
113. A. Sponner, *et al.*, Composition and Hierarchical Organisation of a Spider Silk, *PLoS One*, 2007, **2**, e998.
114. Y. Yokosaki, *et al.*, The integrin alpha 9 beta 1 mediates cell attachment to a non-RGD site in the third fibronectin type III repeat of tenascin, *J. Biol. Chem.*, 1994, **269**, 26691–26696.
115. T. C. Holmes, *et al.*, Extensive neurite outgrowth and active synapse formation on self-assembling peptide scaffolds, *Proc. Natl. Acad. Sci. U. S. A.*, 2000, **97**, 6728–6733.
116. P. B. Harbury, T. Zhang, P. S. Kim and T. Alber, A switch between two-, three-, and four-stranded coiled coils in GCN4 leucine zipper mutants, *Science*, 1993, **262**, 1401–1407.
117. E. K. O'Shea, R. Rutkowski and P. S. Kim, Evidence that the leucine zipper is a coiled coil, *Science*, 1989, **243**, 538–542.
118. F. H. C. Crick, The packing of -helices: simple coiled-coils, *Acta Crystallogr.*, 1953, **6**, 689–697.
119. E. K. O'Shea, K. J. Lumb and P. S. Kim, Peptide "Velcro": Design of a heterodimeric coiled coil, *Curr. Biol.*, 1993, **3**, 658–667.
120. A. N. Lupas and J. Bassler, Coiled Coils - A Model System for the 21st Century, *Trends Biochem. Sci.*, 2017, **42**, 130–140.
121. J. H. Brown, K.-H. Kim, G. Jun, N. J. Greenfield, R. Dominguez, N. Volkmann, S. E. Hitchcock-DeGregori and C. Cohen, Deciphering the design of the tropomyosin molecule, *Proc. Natl. Acad. Sci. U. S. A.*, 2001, **98**, 8496–8501.
122. A. Karabinos, J. Schünemann, M. Meyer, U. Aebi and K. Weber, The Single Nuclear Lamin of Caenorhabditis elegans Forms In Vitro Stable Intermediate Filaments and Paracrystals with a Reduced Axial Periodicity, *J. Mol. Biol.*, 2003, **325**, 241–247.
123. D. Papapostolou, *et al.*, Engineering nanoscale order into a designed protein fiber, *Proc. Natl. Acad. Sci. U. S. A.*, 2007, **104**, 10853–10858.
124. M. G. Ryadnov and D. N. Woolfson, Self-Assembled Templates for Polypeptide Synthesis, *J. Am. Chem. Soc.*, 2007, **129**, 14074–14081.
125. H. Herrmann and U. Aebi, Intermediate Filaments: Structure and Assembly, *Cold Spring Harbor Perspect. Biol.*, 2016, **8**, a018242.
126. H. Herrmann and U. Aebi, Intermediate Filaments: Molecular Structure, Assembly Mechanism, and Integration Into Functionally Distinct Intracellular Scaffolds, *Annu. Rev. Biochem.*, 2004, **73**, 749–789.
127. E. H. Bromley, K. J. Channon, P. J. King, Z. N. Mahmoud, E. F. Banwell, M. F. Butler, M. P. Crump, T. R. Dafforn, M. R. Hicks, J. D. Hirst, A. Rodger and D. N. Woolfson, Assembly pathway of a designed alpha-helical protein fiber, *Biophys. J.*, 2010, **98**, 1668–1676.
128. K. Weber, *et al.*, SF-assemblin, the structural protein of the 2-nm filaments from striated microtubule associated fibers of algal flagellar roots, forms a segmented coiled coil, *J. Cell Biol.*, 1993, **121**, 837–845.

129. E. F. Banwell, E. S. Abelardo, D. J. Adams, M. A. Birchall, A. Corrigan, A. M. Donald, M. Kirkland, L. C. Serpell, M. F. Butler and D. N. Woolfson, *Nat. Mater.*, 2009, **8**, 596–600.
130. E. Fuchs, Intermediate filaments and disease: mutations that cripple cell strength, *J. Cell Biol.*, 1994, **125**, 511–516.
131. H. Bär, N. Mücke, A. Kostareva, G. Sjöberg, U. Aebi and H. Herrmann, Severe muscle disease-causing desmin mutations interfere with *in vitro* filament assembly at distinct stages, *Proc. Natl. Acad. Sci. U. S. A.*, 2005, **102**, 15099–15104.
132. S. A. Potekhin, *et al.*, De novo design of fibrils made of short alpha-helical coiled coil peptides, *Chem. Biol.*, 2001, **8**, 1025–1032.
133. E. De Santis, N. Faruqui, J. E. Noble and M. G. Ryadnov, Exploitable length correlations in peptide nanofibres, *Nanoscale*, 2014, **6**, 11425–11430.
134. F. Wang, O. Gnewou, C. Modlin, L. C. Beltran, C. Xu, Z. Su, P. Juneja, G. Grigoryan, E. H. Egelman and V. P. Conticello, Structural analysis of cross α-helical nanotubes provides insight into the designability of filamentous peptide nanomaterials, *Nat. Commun.*, 2021, **12**, 407.
135. E. De Santis, V. Castelletto and M. G. Ryadnov, Interfacial zippering-up of coiled-coil protein filaments, *Phys. Chem. Chem. Phys.*, 2015, **17**, 31055–31060.
136. H. Shen, J. A. Fallas, E. Lynch, W. Sheffler, B. Parry, N. Jannetty, J. Decarreau, M. Wagenbach, J. J. Vicente, J. Chen, L. Wang, Q. Dowling, G. Oberdorfer, L. Stewart, L. Wordeman, J. De Yoreo, C. Jacobs-Wagner, J. Kollman and D. Baker, De novo design of self-assembling helical protein filaments, *Science*, 2018, **362**, 705–709.
137. D. E. Wagner, *et al.*, Toward the development of peptide nanofilaments and nanoropes as smart materials, *Proc. Natl. Acad. Sci. U. S. A.*, 2005, **102**, 12656–12661.
138. M. Nambiar, L. S. Wang, V. Rotello and J. Chmielewski, Reversible Hierarchical Assembly of Trimeric Coiled-Coil Peptides into Banded Nano- and Microstructures, *J. Am. Chem. Soc.*, 2018, **140**, 13028–13033.
139. K. F. Lechtreck, Analysis of striated fiber formation by recombinant SF-assemblin *in vitro*, *J. Mol. Biol.*, 1998, **279**, 423–438.
140. T. H. Sharp, M. Bruning, J. Mantell, R. B. Sessions, A. R. Thomson, N. R. Zaccai, R. L. Brady, P. Verkade and D. N. Woolfson, Cryo-transmission electron microscopy structure of a gigadalton peptide fiber of *de novo* design, *Proc. Natl. Acad. Sci. U. S. A.*, 2012, **109**, 13266–13271.
141. D. Wu, N. Sinha, J. Lee, B. P. Sutherland, N. I. Halaszynski, Y. Tian, J. Caplan, H. V. Zhang, J. G. Saven, C. J. Kloxin and D. J. Pochan, Polymers with controlled assembly and rigidity made with click-functional peptide bundles, *Nature*, 2019, **574**, 658–662.
142. M. G. Ryadnov and D. N. Woolfson, Engineering the morphology of a self-assembling protein fibre, *Nat. Mater.*, 2003, **2**, 329–332.
143. M. G. Ryadnov and D. N. Woolfson, Introducing Branches into a Self-Assembling Peptide Fiber, *Angew. Chem., Int. Ed.*, 2003, **42**, 3021–3023.

144. M. G. Ryadnov and D. N. Woolfson, MaP Peptides: Programming the Self-Assembly of Peptide-Based Mesoscopic Matrices, *J. Am. Chem. Soc.*, 2005, **127**, 12407–12415.
145. M. Zhou, D. Bentley and I. Ghosh, Helical supramolecules and fibers utilizing leucine zipper displaying dendrimers, *J. Am. Chem. Soc.*, 2004, **126**, 734–735.
146. M. G. Ryadnov, A. Bella, S. Timson and D. N. Woolfson, Modular Design of Peptide Fibrillar Nano- to Microstructures, *J. Am. Chem. Soc.*, 2009, **131**, 13240–13241.
147. A. Bella, S. Ray, M. Shaw and M. G. Ryadnov, Arbitrary self-assembly of peptide extracellular microscopic matrices, *Angew. Chem., Int. Ed. Engl.*, 2012, **51**, 428–431.
148. N. Faruqui, A. Bella, J. Ravi, S. Ray, B. Lamarre and M. G. Ryadnov, Differentially instructive extracellular protein micro-nets, *J. Am. Chem. Soc.*, 2014, **136**, 7889–7898.
149. L. Zajiczek, M. Shaw, N. Faruqui, A. Bella, V. M. Pawar, M. A. Srinivasan and M. G. Ryadnov, Nano-mechanical single-cell sensing of cell-matrix contacts, *Nanoscale*, 2016, **8**, 18105–18112.
150. M. G. Ryadnov and D. N. Woolfson, Fiber Recruiting Peptides: Noncovalent Decoration of an Engineered Protein Scaffold, *J. Am. Chem. Soc.*, 2004, **126**, 7454–7455.
151. G. A. Silva, *et al.*, Selective Differentiation of Neural Progenitor Cells by High-Epitope Density Nanofibers, *Science*, 2004, **303**, 1352–1355.
152. M. P. Hendricks, K. Sato, L. C. Palmer and S. I. Stupp, Supramolecular Assembly of Peptide Amphiphiles, *Acc. Chem. Res.*, 2017, **50**, 2440–2448.
153. J. H. Ortony, C. J. Newcomb, J. B. Matson, L. C. Palmer, P. E. Doan, B. M. Hoffman and S. I. Stupp, Internal dynamics of a supramolecular nanofibre, *Nat. Mater.*, 2014, **13**, 812–816.
154. C. H. Chen, E. L. Hsu and S. I. Stupp, Supramolecular self-assembling peptides to deliver bone morphogenetic proteins for skeletal regeneration, *Bone*, 2020, **141**, 115565.
155. S. Zhou, A. Hokugo, M. McClendon, Z. Zhang, R. Bakshi, L. Wang, L. A. Segovia, K. Rezzadeh, S. I. Stupp and R. Jarrahy, Bioactive peptide amphiphile nanofiber gels enhance burn wound healing, *Burns*, 2019, **45**, 1112–1121.
156. E. Sleep, B. D. Cosgrove, M. T. McClendon, A. T. Preslar, C. H. Chen, M. H. Sangji, C. M. R. Pérez, R. D. Haynes, T. J. Meade, H. M. Blau and S. I. Stupp, Injectable biomimetic liquid crystalline scaffolds enhance muscle stem cell transplantation, *Proc. Natl. Acad. Sci. U. S. A.*, 2017, **114**, E7919–E7928.
157. C. E. Morgan, A. W. Dombrowski, C. M. Rubert Pérez, E. S. Bahnson, N. D. Tsihlis, W. Jiang, Q. Jiang, J. M. Vercammen, V. S. Prakash, T. A. Pritts, S. I. Stupp and M. R. Kibbe, Tissue-factor targeted peptide amphiphile nanofibers as an injectable therapy to control hemorrhage, *ACS Nano*, 2016, **10**, 899–909.

158. F. Gelain, D. Silva, A. Caprini, F. Taraballi, A. Natalello, O. Villa, K. T. Nam, R. N. Zuckermann, S. M. Doglia and A. Vescovi, BMHP1-derived self-assembling peptides: hierarchically assembled structures with self-healing propensity and potential for tissue engineering applications, *ACS Nano*, 2011, **5**, 1845–1859.
159. X. Ding, H. Zhao, Y. Li, A. L. Lee, Z. Li, M. Fu, C. Li, Y. Y. Yang and P. Yuan, Synthetic peptide hydrogels as 3D scaffolds for tissue engineering, *Adv. Drug Delivery Rev.*, 2020, **160**, 78–104.
160. N. Stephanopoulos, J. H. Ortony and S. I. Stupp, Self-Assembly for the Synthesis of Functional Biomaterials, *Acta Mater.*, 2013, **61**, 912–930.
161. D. N. Woolfson and Z. N. Mahmoud, More than just bare scaffolds: towards multi-component and decorated fibrous biomaterials, *Chem. Soc. Rev.*, 2010, **39**, 3464–3479.
162. A. Y. Wang, X. Mo, C. S. Chen and S. M. Yu, Facile Modification of Collagen Directed by Collagen Mimetic Peptides, *J. Am. Chem. Soc.*, 2005, **127**, 4130–4131.
163. J. Hwang, Y. Huang, T. J. Burwell, N. C. Peterson, J. Connor, S. J. Weiss, S. M. Yu and Y. Li, In Situ Imaging of Tissue Remodeling with Collagen Hybridizing Peptides, *ACS Nano*, 2018, **11**, 9825–9835.
164. J. H. Collier, *et al.*, Thermally and photochemically triggered self-assembly of peptide hydrogels, *J. Am. Chem. Soc.*, 2001, **123**, 9463–9464.
165. N. M. Mascarenhas and S. Gosavi, Understanding protein domain-swapping using structure-based models of protein folding, *Prog. Biophys. Mol. Biol.*, 2017, **128**, 113–120.
166. M. S. Lamm, K. Rajagopal, J. P. Schneider and D. J. Pochan, Laminated Morphology of Nontwisting beta-Sheet Fibrils Constructed *via* Peptide Self-Assembly, *J. Am. Chem. Soc.*, 2005, **127**, 16692–16700.
167. R. P. Nagarkar, R. A. Hule, D. J. Pochan and J. P. Schneider, Domain swapping in materials design, *Biopolymers*, 2010, **94**, 141–155.
168. R. P. Nagarkar, R. A. Hule, D. J. Pochan and J. P. Schneider, *J. Am. Chem. Soc.*, 2008, **130**, 4466–4474.
169. D. J. Smith, G. A. Brat, S. H. Medina, D. Tong, Y. Huang, J. Grahammer, G. J. Furtmüller, B. C. Oh, K. J. Nagy-Smith, P. Walczak, G. Brandacher and J. P. Schneider, A multiphase transitioning peptide hydrogel for suturing ultrasmall vessels, *Nat. Nanotechnol.*, 2016, **11**, 95–102.
170. M. Reches and E. Gazit, Casting Metal Nanowires Within Discrete Self-Assembled Peptide Nanotubes, *Science*, 2003, **300**, 625–627.
171. C. H. Görbitz, Nanotube Formation by Hydrophobic Dipeptides, *Chemistry*, 2001, **7**, 5153–5159.
172. C. H. Görbitz, An NH3+...phenyl interaction in L-phenylalanyl-L-valine, *Acta Crystallogr. C*, 2000, **56**, 1496–1498.
173. Y. Zhang, H. Gu, Z. Yang and B. Xu, Supramolecular Hydrogels Respond to Ligand-Receptor Interaction, *J. Am. Chem. Soc.*, 2003, **125**, 13680–13681.

174. V. Jayawarna, *et al.*, Nanostructured Hydrogels for Three-Dimensional Cell Culture Through Self-Assembly of Fluorenylmethoxycarbonyl-Dipeptides, *Adv. Mater.*, 2006, **18**, 611–614.
175. S. Fleming and R. V. Ulijn, Design of nanostructures based on aromatic peptide amphiphiles, *Chem. Soc. Rev.*, 2014, **43**, 8150–8177.
176. A. M. Smith, *et al.*, Fmoc-Diphenylalanine Self Assembles to a Hydrogel *via* a Novel Architecture Based on pi-pi Interlocked beta-Sheets, *Adv. Mater.*, 2008, **20**, 37–41.
177. C. G. Pappas, R. Shafi, I. R. Sasselli, H. Siccardi, T. Wang, V. Narang, R. Abzalimov, N. Wijerathne and R. V. Ulijn, Dynamic peptide libraries for the discovery of supramolecular nanomaterials, *Nat. Nanotechnol.*, 2016, **11**, 960–967.
178. X. Du, J. Zhou, J. Shi and B. Xu, Supramolecular Hydrogelators and Hydrogels: From Soft Matter to Molecular Biomaterials, *Chem. Rev.*, 2015, **115**, 13165–13307.
179. R. M. Nerem, Tissue Engineering: The Hope, the Hype, and the Future, *Tissue Eng.*, 2006, **12**, 1143–1150.
180. S. Talebian, M. Mehrali, N. Taebnia, C. P. Pennisi, F. B. Kadumudi, J. Foroughi, M. Hasany, M. Nikkhah, M. Akbari, G. Orive and A. Dolatshahi-Pirouz, Self-Healing Hydrogels: The Next Paradigm Shift in Tissue Engineering?, *Adv. Sci.*, 2019, **6**, 1801664.
181. J. J. Marler, J. Upton, R. Langer and J. P. Vacanti, Transplantation of cells in matrices for tissue regeneration, *Adv. Drug Delivery Rev.*, 1998, **33**, 165–182.
182. J. R. Fuchs, B. A. Nasseri and J. P. Vacanti, Tissue engineering: a 21st century solution to surgical reconstruction, *Ann. Thorac. Surg.*, 2001, **72**, 577–591.
183. R. Langer and J. Vacanti, Advances in tissue engineering, *J. Pediatr. Surg.*, 2016, **51**, 8–12.
184. T. Hoffman, A. Khademhosseini and R. Langer, Chasing the Paradigm: Clinical Translation of 25 Years of Tissue Engineering, *Tissue Eng., Part A*, 2019, **25**, 679–687.
185. R. M. Nerem, Tissue engineering: confronting the transplantation crisis, *Adv. Exp. Med. Biol.*, 2003, **534**, 1–9.
186. D. L. Stocum, *Regenerative Biology and Medicine*, Elsevier, Oxford, 2006.
187. A. Vats, N. S. Tolley, A. E. Bishop and J. M. Polak, Embryonic stem cells and tissue engineering: delivering stem cells to the clinic, *J. R. Soc. Med.*, 2005, **98**, 346–350.
188. K. A. Moore and I. R. Lemischka, Stem cells and their niches, *Science*, 2006, **311**, 1880–1885.
189. I. L. Weissman, Stem Cells: Units of Development, Units of Regeneration, and Units in Evolution, *Cell*, 2000, **100**, 157–168.
190. H. M. Blau, T. R. Brazelton and J. M. Weimann, The Evolving Concept of a Stem Cell: Entity or Function?, *Cell*, 2001, **105**, 829–841.
191. P. Bianco and P. G. Robey, Stem cells in tissue engineering, *Nature*, 2001, **414**, 118–121.

192. A. E. Bishop, L. D. Buttery and J. M. Polak, Embryonic stem cells, *J. Pathol.*, 2002, **197**, 424–429.
193. V. M. Weaver, *et al.*, Reversion of the malignant phenotype of human breast cells in three-dimensional culture and *in vivo* by integrin blocking antibodies, *J. Cell Biol.*, 1997, **137**, 231–245.
194. H. Peinado, H. Zhang, I. R. Matei, B. Costa-Silva, A. Hoshino, G. Rodrigues, B. Psaila, R. N. Kaplan, J. F. Bromberg, Y. Kang, M. J. Bissell, T. R. Cox, A. J. Giaccia, J. T. Erler, S. Hiratsuka, C. M. Ghajar and D. Lyden, Pre-metastatic niches: organ-specific homes for metastases, *Nat. Rev. Cancer*, 2017, **17**, 302–317.
195. O. R. Tonti, H. Larson, S. N. Lipp, C. M. Luetkemeyer, M. Makam, D. Vargas, S. M. Wilcox and S. Calve, Tissue-specific parameters for the design of ECM-mimetic biomaterials, *Acta Biomater.*, 2021, **132**, 83–102.
196. E. Cukierman, R. Pankov and K. M. Yamada, Cell interactions with three-dimensional matrices, *Curr. Opin. Cell Biol.*, 2002, **14**, 633–640.
197. E. Cukierman, R. Pankov, D. R. Stevens and K. M. Yamada, Taking Cell-Matrix Adhesions to the Third Dimension, *Science*, 2001, **294**, 1708–1712.
198. D. A. C. Walma and K. M. Yamada, The extracellular matrix in development, *Development*, 2020, **147**, dev175596.
199. K. M. Yamada and M. Sixt, Mechanisms of 3D cell migration, *Nat. Rev. Mol. Cell Biol.*, 2019, **20**, 738–752.
200. G. Chan and D. J. Mooney, New materials for tissue engineering: towards greater control over the biological response, *Trends Biotechnol.*, 2008, **26**, 382–392.
201. S. Raghavan and C. S. Chen, Micropatterned Environments in Cell Biology, *Adv. Mater.*, 2004, **16**, 1303–1313.
202. A. Folch and M. Toner, Microengineering of cellular interactions, *Annu. Rev. Biomed. Eng.*, 2000, **2**, 227–256.
203. C. M. Nelson, M. M. VanDuijn, J. L. Inman, D. A. Fletcher and M. J. Bissell, Tissue Geometry Determines Sites of Mammary Branching Morphogenesis in Organotypic Cultures, *Science*, 2006, **314**, 298–300.
204. J. J. Tomasek, G. Gabbiani, B. Hinz, C. Chaponnier and R. A. Brown, Myofibroblasts and mechano-regulation of connective tissue remodelling, *Nat. Rev. Mol. Cell Biol.*, 2002, **3**, 349–363.
205. E. Tamariz and F. Grinnell, Modulation of Fibroblast Morphology and Adhesion during Collagen Matrix Remodeling, *Mol. Biol. Cell*, 2002, **13**, 3915–3929.
206. B. M. Gillette, *et al.*, In situ collagen assembly for integrating microfabricated three-dimensional cell-seeded matrices, *Nat. Mater.*, 2008, **7**, 636–640.
207. R. Zeiser, A. Beilhack and R. S. Negrin, Acute graft-*versus*-host disease-challenge for a broader application of allogeneic hematopoietic cell transplantation, *Curr. Stem Cell Res. Ther.*, 2006, **1**, 203–212.

208. H. Petite, *et al.*, Tissue-engineered bone regeneration, *Nat. Biotechnol.*, 2000, **18**, 959–963.
209. G. E. Holt, J. L. Halpern, T. T. Dovan, D. Hamming and H. S. Schwartz, Evolution of an *in vivo* bioreactor, *J. Orthop. Res.*, 2005, **23**, 916–923.
210. H. J. Kong and D. J. Mooney, Microenvironmental regulation of biomacromolecular therapies, *Nat. Rev. Drug Discovery*, 2007, **6**, 455–463.
211. R. W. Sands and D. J. Mooney, Polymers to direct cell fate by controlling the microenvironment, *Curr. Opin. Biotechnol.*, 2007, **18**, 448–453.
212. M. M. Stevens, *et al.*, In vivo engineering of organs: The bone bioreactor, *Proc. Natl. Acad. Sci. U. S. A.*, 2005, **102**, 11450–11455.
213. R. P. Marini, M. M. Stevens, R. Langer and V. P. Shastri, Hydraulic Elevation of the Periosteum: A Novel Technique for Periosteal Harvest, *J. Investig. Surg.*, 2004, **17**, 229–233.
214. C. Fischbach, *et al.*, Engineering tumors with 3D scaffolds, *Nat. Methods*, 2007, **4**, 855–860.

CHAPTER 5

Concluding Remarks

The process of molecular self-assembly is central to biological organization. With the wealth of monomeric macromolecular structures, such as proteins and DNA, Nature prefers spontaneous non-covalent associations of protomer subunits over complex covalent architectures to build functional nano-to microscopic objects of cells and tissues.

Synthetic designs demonstrate the same preference, which is also rooted in molecular synthesis being reciprocal and complementary to self-assembly. For instance, ribosomes synthesize proteins from amino acids but themselves assemble from proteins. Biological assemblies in most, if not all, cases must be capable of undergoing stimuli-responsive assembly–disassembly cycles. This intrinsic property of biology is provided by molecular encoding – a structure–function rationale for biofunctional nanoscale designs.

That said, self-assembly can be viewed as a programming language for generating hierarchical structures at the nanoscale. Such a language can deal with general operators such as the hydrophobic effect or van der Waals interactions, but also permits more definitive coding interrelations allowing for specific assembly modes, including those that are atypical for natural systems. Thus, the complexity and refinement of the resulting structures correlate strongly with the level at which language is applied and understood.

Our ability to predict and prescribe desired assemblies is improving, and so is the repertoire of available instruments and measurement approaches, with the main efforts being focused on compiling the toolbox of molecular motifs instructive at the nanoscale. The building of such an experimental vocabulary is an empirical process of trial and error, which helps to reveal and select new motifs and associated processes.

Nanoscience & Nanotechnology Series No. 53
Bionanodesign: Old Forms for New Functions, 2nd Edition
By Maxim Ryadnov

Published by the Royal Society of Chemistry, www.rsc.org

Most of all, however, mastering the language of nanoscale design requires structuring – formulating a systematic, general matrix linking molecular patterns and periodicities with desired nanostructures and *vice versa*. Most design approaches reported thus far appear to be specific to individual research teams or groups of teams and present a series of implicitly related "dialects" rather than one common language. We know, for example, that desired DNA sequences can be written to programme complex structures and unknown DNA or RNA sequences can be divided and read to a single nucleotide. But we hardly know how and why some folds of nucleic acids template reproducible nucleoprotein complexes and others do not, and if and when the mutation of a single nucleotide can compromise an entire nanostructure. In this regard, current theories are componential but not necessarily interchangeable. This may explain why the search for effective self-assembly motifs has mainly been concerned with attempts to address particular design challenges, be these structural or functional. This tendency prompts a pragmatic need to systematize the established self-assembly phenomena into a universal protocol.

This is challenging but also seems unavoidable. Learning from experience is probably the shortest route to formulating relationships between basic design principles and self-assembly motifs. There exist three correlative characteristics to consider.

One deals with building blocks and the cooperativity of their interactions upon assembly into higher order structures. This characteristic is set at the level of primary motifs, which define the assembly type and are subject to responses to environmental stimuli, which favour or disfavour the formation of a desired nanostructure form.

Another characteristic perceives the responsive properties of the assembled structures in terms of reversibility and motion. These depend on the inter- and intramolecular dynamics of the interactions impacting on the compatibility of the structures with their environment. For example, if reversibility is a prerequisite for all assembly steps, hierarchical organization is predominantly concerned with local molecular adjustments within the assembled structure, whereas motion relates to the precursor and resultant forms reaching an equilibrium.

The main distinction of the third characteristic is how assembly errors and variations relate to the multiplicity of the interactions. These may include defects at various assembly stages, multiplication *versus* increasing complexity and a means of removal or selection. In natural systems, such a problem is addressed by various proofreading mechanisms. Artificial systems are short of these and lack compensatory measures to rectify imperfections. Therefore, this characteristic is positively discriminative, and can be viewed from the perspective of variational competitions to reveal factors that limit imperfections.

The interest in the design of functional nanoscale systems is tremendous and appears to lean towards formulating reproducible and systemic protocols for experimental designs. This choice feeds into the emerging challenge of commercialization with all its associated attributes of stability, manufacturing and scale-up for applications. This might not be the final destination for bionanodesign, but a connecting bridge between trial-and-error approaches serving to identify autonomously operating systems and the deciphered language of molecular self-assembly.

Subject Index

Figures are indicated by an italic *f* following the page number.